"In this exploration of how the U.S. media covers the environment and environmental justice issues, Moore offers an insightful and compelling discussion of tensions between commercialism, journalism, and coverage of key issues that should be presented as relevant to us all. Through her examinations of our corporate-owned media, Moore presents a timely and important 'Study in Power,' a meticulous, unwavering, and well-researched eye on structural weaknesses in how we receive news of our ever-increasingly vulnerable world."

— *Rena Priest, Lhaq' te' mish author,* Patriarchy Blues

"Ellen Moore provides a brilliant analysis of media coverage of the 2016 Dakota Access Pipeline protests by the Standing Rock Sioux Tribe and the 'water protectors' who were its supporters. Most relevantly, she documents the general failure of modern journalism to adequately cover environmental issues and the concerns of Indigenous peoples. In this effort, her work represents the very finest in communications scholarship."

— *Michael L. Lawson, author of* Dammed Indians Revisited: The Continuing History of the Pick-Sloan Plan and the Missouri River Sioux

"Moore's penetrating study of the media's response to the Standing Rock Sioux Tribe's bold protests against the Dakota Access Pipeline is ultimately a profound indictment of the Fourth Estate's unwillingness to check the greed of the fossil fuel industry or to inspire moral courage in our political leaders. Moving fluidly between first-hand accounts of Sioux leaders and illuminating lessons from the literatures on environmental justice and the political economy of the media industry, Moore reveals how the Standing Rock controversy refracts the media's long-standing reluctance to call out environmental abuses and to shine a bright light on environmental injustices."

— *Heidi M. Hurd, Ross and Helen Workman Chair in Law and Professor of Philosophy at the University of Illinois*

"Engaging and informative, this book uses the lens of media studies to expertly demonstrate how questions of environmental justice, social justice and environmental ethics are tied together in the NODAPL movement. It is a must-read for academics or general readers who are interested in how media coverage can inform our opinions and values and frame agendas in ways that have far-reaching consequences."

— *Jane Compson, Associate Professor, School of Interdisciplinary Arts and Sciences, University of Washington Tacoma*

"Both an engaging, respectful discourse with Indigenous voices and an eviscerating critique of media's complicity in environmental degradation."

— *Danica Sterud Miller (Puyallup), Assistant Professor of American Indian Studies*

"Moore expertly weaves through the confluence of media coverage, Big Oil, and Indigenous activism with considerable poise, in turn revealing a timely, eye-opening, and personal telling on the complexities surrounding the Dakota Access Pipeline protest and its representation throughout various media outlets."

— *Miguel Douglas, Editor-in-chief of* American Indian Republic

JOURNALISM, POLITICS, AND THE DAKOTA ACCESS PIPELINE

This book explores tensions surrounding news media coverage of Indigenous environmental justice issues, identifying them as a fruitful lens through which to examine the political economy of journalism, American history, human rights, and contemporary U.S. politics.

The book begins by evaluating contemporary American journalism through the lens of "deep media," focusing especially on the relationship between the drive for profit, professional journalism, and coverage of environmental justice issues. It then presents the results of a framing analysis of the Standing Rock movement (#NODAPL) coverage by news outlets in the USA and Canada. These findings are complemented by interviews with the Standing Rock Sioux Tribe, whose members provided their perspectives on the media and the pipeline. The discussion expands by considering the findings in light of current U.S. politics, including a Trump presidency that employs "law and order" rhetoric regarding people of color and that often subjects environmental issues to an economic "cost-benefit" analysis. The book concludes by considering the role of social media in the era of "Big Oil" and growing Indigenous resistance and power.

Examining the complex interplay between social media, traditional journalism, and environmental justice issues, *Journalism, Politics, and the Dakota Access Pipeline: Standing Rock and the Framing of Injustice* will be of great interest to students and scholars of environmental communication, critical political economy, and journalism studies more broadly.

Ellen Moore is a Senior Lecturer at the University of Washington Tacoma, USA.

ROUTLEDGE STUDIES IN ENVIRONMENTAL COMMUNICATION AND MEDIA

For more information about this series, please visit: https://www.routledge.com/Routledge-Studies-in-Environmental-Communication-and-Media/book-series/RSECM

JOURNALISM, POLITICS, AND THE DAKOTA ACCESS PIPELINE

Standing Rock and the Framing of Injustice

Ellen Moore

Routledge
Taylor & Francis Group
LONDON AND NEW YORK

earthscan
from Routledge

First published 2019
by Routledge
2 Park Square, Milton Park, Abingdon, Oxon OX14 4RN

and by Routledge
52 Vanderbilt Avenue, New York, NY 10017

Routledge is an imprint of the Taylor & Francis Group, an informa business

British Library Cataloguing-in-Publication Data
A catalogue record for this book is available from the British Library

Library of Congress Cataloging-in-Publication Data
A catalog record has been requested for this book

ISBN: 978-0-8153-9909-4 (hbk)
ISBN: 978-0-8153-9941-4 (pbk)
ISBN: 978-1-351-17176-2 (ebk)

Typeset in Bembo
by Deanta Global Publishing Services, Chennai, India

CONTENTS

PREFACE

Background and language

In the beginning of *The New Media Nation: Indigenous Peoples and Global Communication*, Valerie Alia contends that, "This is not my story." What she means by this is that, as a white Eastern European-North American, she may research and chronicle the Indigenous experience when it comes to the media, but that ultimately, she does not own the story. Her important recognition of this is similar to my own – namely, that as a white American of Scottish ancestry, writing on the events that occurred at Standing Rock in the autumn of 2016 (when the resistance to the Dakota Access Pipeline reached its zenith), this is a story about the relationship between the Standing Rock Sioux Tribe and the media, but clearly it is not my narrative. While throughout the book I examine what is commonly referred to as the #NODAPL movement through the lens of new and legacy media, I do not claim to speak for the Tribe; instead, the Tribe's voice emerges during my interviews with them as chronicled in Chapter Seven. In addition, the Tribe has read and has provided feedback on this book prior to its publication.

In many ways, writing this book was a lesson in cultural humility, a study in privilege, and a recognition of resilience. Being able to work with the Tribe was of immeasurable value for this book but also a great responsibility, for Indigenous groups have been misrepresented and stereotyped by the media and in academia for centuries. Thus, it was important to earn their trust and get their story regarding the #NODAPL movement, the pipeline, and the media right. In addition, after my interviews with several members of the Tribe, I could not see the world the same way again. I remember one example very clearly: after completing interviews with the Tribe on an evening in late April, 2018, I drove to get dinner at the Tribe's restaurant and casino (Prairie Knights). When I took out a $20 bill to pay for dinner I stared at it in shock: the Tribe had just talked with me about their perspective on former President Andrew Jackson, who was known for signing the injurious *Indian Removal Act* as well as many other acts of violence against many Native American Tribes. Although I had known about his history with Indigenous groups prior to meeting with the Tribe, I always had the privilege of seeing "just" a $20 bill. Held in my hands after the interview the bill took on new meaning, for I realized it would act as a constant reminder to the Tribe of a long, troubled history with the U.S. government, one marked by significant loss and injustice.

Meeting with the Tribe also taught me a lesson in *resilience*, for when I queried them regarding how they felt about the current political situation with President Trump (who has indicated he'd like their sovereign status to be removed, among other things), their answer surprised me. Instead of angry retorts or broad judgments of him, their answer was simple: *We lived through Andrew Jackson, we lived through Lincoln, we'll live through Trump. So, you know, we'll always be here.* Their quiet strength introduced to me a new perspective on resilience also related to my privilege: while I, as a white, middle-class progressive, felt that the sky was precipitously "falling" with a Trump presidency (due to reasons related to human rights, environmental degradation, animal welfare, consumer protections, and more) for the Tribe, it was another challenge that, while serious, they were strong enough to survive and overcome.

Thus, it is accurate to say that the Tribe gently and patiently took me to school on many issues, fundamentally changing my perspective by gaining some of theirs. For this reason I am grateful for the Tribe's time and contribution to this book. It is also for this reason that the book is dedicated to them.

Notes on language and imagery used in the book

It is important to note several patterns in the use of language in this book. Language, as a symbol in itself, can hold great significance, and thus in this book I have chosen my words carefully. In these next sections I make clear how and why I chose to use certain images, words, and phrases.

Describing the resistance movement at Standing Rock in 2016

I refer to the resistance to the Dakota Access Pipeline that occurred in 2016 and 2017 in a few, interchangeable ways. Most commonly I refer to the movement as "#NODAPL" (recognizing that the #NDAPL hashtag also has been used), "the resistance," "the demonstrations," "the gathering," and, far less frequently, "the protests." The reason "protest" is used less often is that the Tribe felt that it had taken on a negative connotation and thus I tried to avoid this phrasing when and where possible. The "gathering" seems a common phrase used by the Tribe to refer to the coming together of many Tribes on an international scale. All terms, when used in this book, refer to the same thing: the mass resistance, prayer, and ceremony that occurred in the fight against the pipeline.

Describing those who participated in #NODAPL

Those who gathered at Standing Rock in 2016 have been identified in myriad ways, including as "protesters," "rioters," "water protectors," and "resistors." I studiously avoided use of the term "rioter," tried to use "protesters" as minimally as possible, and often employed "resistors" or "demonstrators" as a general descriptive term.

Capitalization of words to recognize sovereignty of Indigenous groups

As I and some other scholars have done in previous academic works, certain words are capitalized as a way to visually represent, on the page, the sovereignty of Indigenous nations. Thus, the reader will encounter capitalization of words like "Indigenous" and "Tribes" as a rule as a consistent reminder of the sovereign status of the Standing Rock Sioux and other Indigenous groups.

Use of language to describe Indigenous groups

Throughout the book I use "Indigenous groups" and "Native Americans" as broad designators, with the former preferred and thus employed more often than the latter. Although some Indigenous scholars use designators like "American Indian" or "Indian," I tended to avoid use of those terms because they are used in specific contexts and contain their own political meaning. When writing about the Standing Rock Sioux Tribe, my tendency is to write out the full name, or use the word "Tribe" when it is clear to whom I am referring.

Choosing a cover image for the book

An unexpected challenge came in the form of choosing a cover image for the book. Many of the images that I originally selected from "stock" photos taken of the movement (which can be purchased from many image repositories like Alamy and Getty) contained clearly recognizable Indigenous faces. While these images were compelling, using them felt potentially exploitative because the people in them might not want to be seen as representing or endorsing the book. Other images, ones that showed white people supporting the movement in various cities around the U.S., also were problematic: while the movement did indeed extend beyond Standing Rock to gain global support, this was not a white-led movement. As a result, I worked with the Standing Rock Sioux Tribe to choose an image that was both relevant to the movement as well as respectful to those involved in it. The cover image you see now is the result of that collaboration, depicting water protectors and military veterans marching together in the snow as winter sets in on December 5, 2016, the day after the U.S. Army Corps temporarily blocked construction of the pipeline.

ACKNOWLEDGMENTS

As is the case with most books, this was not a solo effort, and there were many people who made the creation of this work possible. First, I would like to thank the Standing Rock Sioux Tribe for providing me with the gift of their time: without them this book would be missing their important perspectives, words, and beliefs. Their patience and generosity made this book what it is.

Related to the above recognition is my gratitude to the School of Interdisciplinary Arts & Sciences at the University of Washington Tacoma. The funding I received from them made my trip to the Standing Rock Reservation possible to conduct the interviews. This financial contribution came in the form of the *Scholarship and Teaching Fund*, provided to faculty for their teaching and research needs as they arise.

Another essential supporter of this book is my husband Paul, who somehow graciously accepted that I started writing this book before I was finished with my first book! His constant gift of his own time meant that this book could be completed with as little stress as possible. Here I'd like to say "thank you" to him while promising in earnest not to start another book in the next year.

A special note of thanks goes to Dr. Danica Sterud Miller, who runs the Lushootseed Language Institute offered as part of the Puyallup Language Program in partnership with the University of Washington Tacoma. She studiously read every single chapter and provided important feedback, even on the often overlooked endnotes! I am grateful for her time and effort. I also am glad to recognize Anna Bean, Puyallup Tribe Councilmember and spokesperson for the Water Warriors, who took time out of her busy schedule to speak to me about the local movement (#NOLNG253) resisting the liquid natural gas plant placed on the Tribe's ancestral homelands.

Many other scholars and journalists took their time to read this book and provide comments and feedback. Dr. Doreen Marchionni, editor-at-large in the Seattle region, read my framing analysis of newspapers within the U.S. and provided much-needed, insightful comments from a journalist and editor's perspective. Similarly, Joshua Meyrowitz, Professor Emeritus in the Department of Communication at the University of New Hampshire, took time to read the book draft and provide unique and productive suggestions – even while traveling and doing his own research. Kylie Lanthorn, former student and now co-author, friend, and graduate student at the University of Massachusetts Amherst, provided invaluable feedback on the framing analysis. This book is much stronger because of these supportive suggestions and critiques and I am grateful for them.

As always, my family is a needed and appreciated form of support. Many thanks for all of your love and kind words during this process. Your support makes it all worthwhile.

1

INTRODUCTION

Media coverage of Standing Rock and the Dakota Access Pipeline: the couple from Bismarck

While on sabbatical in the winter of 2017 with my family, I met a couple from Bismarck, North Dakota who changed the way I thought about the conflicts around the Dakota Access Pipeline. For many months, the ongoing struggles over the pipeline in North Dakota had been on my mind, and I was not alone. A video – published in September 2016 by journalist Amy Goodman with *Democracy Now!* – had captivated media audiences worldwide. The clip, which depicted pipeline protesters (most of whom were Indigenous) being bitten by dogs and attacked with pepper spray went viral, and the global attention it received successfully pushed mainstream media outlets to end their news blackout on the conflict and cover it in earnest. Thus, in the autumn of 2016 I found myself, like many others, closely following the efforts of the Standing Rock Sioux Tribe to stop an oil pipeline from being built directly upstream from their main source of drinking water. At issue was the fact that the pipeline was originally scheduled to cross the Missouri River north of the town of Bismarck, North Dakota (Dalrymple, 2016); reports began to surface, however, that the U.S. Army Corps of Engineers determined that the pipeline might endanger municipal water supplies and subsequently moved the pipeline downstream of the town but directly upstream of the Standing Rock reservation (Dalrymple, 2016; Thorbecke, 2016).

As the weather grew cooler in late autumn 2016, I had watched the increasingly tense struggle through the lens of mainstream news media but also, increasingly, through Facebook and Twitter postings of people who were on the scene. Then more disturbing images emerged: violent encounters between demonstrators and local law enforcement and "water protectors" being led away in handcuffs. As the *New York Times* reported, the Morton County sheriff's office released a video of its use of water cannons on protesters in freezing weather, seemingly in hopes of gaining support for law enforcement's cause, but many perceived that imagery as emblematic of a David and Goliath struggle, one in which Indigenous protesters represented the vulnerable party.

When I was done teaching my classes, I planned to travel to North Dakota; however, a few days prior to my departure in early December, President Obama signed an executive order instructing the U.S. Army Corps to deny pipeline corporation Energy Transfer Partners the permit needed to complete the project. The Tribe and their supporters breathed a cautious sigh of relief as the then chairman of Standing Rock Dave Archambault II asked the majority of protesters to leave. There were others – especially those in North Dakota for whom the pipeline meant an economic boon – however, who were upset that the pipeline construction was at a standstill.

It was during this temporary hiatus of both pipeline construction and protest that I found myself sitting next to a white, middle-aged couple from Bismarck while having drinks at a Hawaiian hotel in late January of 2017. Intrigued that they lived so close to the site of the protests, I engaged them in conversation to learn more: what had actually happened over there, and how did they feel about it? Their answers, on many levels, surprised me. First, they told me that they knew that the pipeline protesters weren't really from the Standing Rock Sioux Tribe but in fact were all violent out-of-towners simply looking to cause trouble. To them, the protest stemmed *not* from deep concerns held by the Tribe regarding their water supply but was instead more politically motivated. When I asked the couple why they thought the pipeline had been moved from where it was originally planned (upstream from Bismarck) to be directly upstream from the Standing Rock reservation, they told me confidently that it was a lie: "They were never going to put it upstream of Bismarck. That's fake news," the husband told me. He then advised me to be careful with my news sources so as to not be misinformed. I asked what news sources he thought were reliable – that is, who would be good to trust? To this he demurred, responding somewhat vaguely: "Oh, we know where to look." They had to catch their bus to the airport, but as they got up to leave I asked the man what he did for a living. His response: "I work for a local oil company."

This conversation was provocative for myriad reasons relating to this book, not the least of which is the reference to the lack of trust in the mainstream news media, or what is so commonly referred to in the general vernacular now as "fake news." The man from Bismarck, like so many news audience members now, felt a fair amount of mistrust towards the media and believed that he needed to be cautious because of this. His statement also is interesting because he seemed to refer to any news he didn't agree with as inauthentic, which mimics the stance of U.S. President Trump and his supporters (who often identify mainstream media outlets that have criticized him as "fake"). However, it is important to note that many people of the other side of the political spectrum are increasingly using the phrase "fake news" in this way as well (Oremus, 2016), making it less a partisan issue and more akin to a broader social phenomenon. The fact that the original pipeline location was confirmed through an examination of U.S. Army

Corps documents and reported on by local, professional journalism outlets did not seem to matter to the man from Bismarck when it came to how he *felt*, which cannot be overstated in its significance regarding contemporary perceptions of the news media (as well as politics) in the U.S.[1]

Another point of intersection between this book and the conversation with the Bismarck couple relates to the environmental justice concerns of Indigenous groups around the world when it comes to the fossil fuel industry. Hydraulic fracturing ("fracking") in the U.S. and Canada, as well as significant changes made by OPEC, have resulted in a glut of oil and gas ready for consumption and export: "the US shale oil boom took off in just a few years and now adds millions of barrels per day to the global supply" (Thomas, 2016). Those millions of extra barrels a day places pressure on the U.S. shale industry to transport its product to market, which then creates significant pressure on local communities across the U.S. to participate in fossil fuel export and transportation projects. As one example, in the U.S. Pacific Northwest several fights have taken place over two different fossil fuel projects that have engaged both residential communities as well as local Indigenous groups like the Puyallup and Lummi Tribes. The Lummi Nation was successful in fighting off a proposed coal terminal – the largest ever proposed – at Cherry Point. The decision came when the U.S. Army Corps of Engineers refused to grant a permit, with the decision based on a respect for historical fishing rights guaranteed by treaty. For the Puyallup Tribe, the fight to stop the completion of a large liquid natural gas plant on their ancestral lands is ongoing as of this writing.

Third, the comments from the Bismarck couple about the inauthenticity of the protests – that is, they did not believe that members of the Standing Rock Sioux Tribe were protesting but instead were violent out-of-towners with no real vested interest beyond creating chaos – seem to discredit the Tribe's partnership in the struggle. In other words, it implied that their environmental concerns were not legitimate enough for such a large-scale protest and widespread national concern, and thus seemed to be a way to discredit or minimize Indigenous rights and environmental concerns.

Finally, the couple from Bismarck were themselves heavily engaged with the fossil fuel industry, seeing it as the "business as usual" commercial enterprise that has indeed grown the local, national, and global economy. We are, as former U.S. president George Bush famously noted in 2006, truly "addicted to oil," and as a result our global economic system is inextricably predicated on fossil fuels (Newman et al., 2017; Morris, 2015; Princen et al., 2015; Jackson, 2010; Juhasz, 2008; Sampson, 1975). In the U.S. (like many places around the world) we rely on fossil fuels for transportation of goods and people, agriculture and food systems, and myriad other facets of contemporary human life. Wesley Jackson (2010) provides the imagery of *scaffolding* as a way to understand this reliance, noting that our human societies are fundamentally predicated on fossil fuel consumption as a way to produce energy: "What we failed to appreciate is how quickly the 'scaffolding' of civilization became so elaborate and so energy intensive" (pp. 11–12). Cusick (2013) paints a picture of this based on skyrocketing U.S. oil consumption:

> For the first time in 2012, global gas production exceeded 3 billion metric tons, marking the third consecutive year of both rising production and consumption, according to the report.... Oil, too, has seen a surge in production in the United States.... In 2012, the United States produced oil at record levels and is expected to overtake Russia this year as the world's largest producer of oil and natural gas combined.[2]

The problem with this heavy dependence on fossil fuel based energy is manifold but can be roughly reduced to three broad concerns: 1) when oil becomes prohibitively expensive and/or scarce, it will exert a significant and detrimental impact on our entire agricultural and economic system (Jackson, 2010; Morris, 2015, 2) our high fossil fuel usage (which continues to rise) occurs at a time of heightened global concern about climate change; and 3) extractive processes and fossil fuel industrial projects often negatively impact Indigenous groups and other communities of color (Princen et al., 2015; Klein, 2014; Bullard & Wright, 2012).

Framed by these concerns, it is clear that while the Standing Rock Sioux Tribe found themselves in a very specific fight against one pipeline, it is impossible to ignore that the struggle takes place within a larger nation that is highly dependent on (as well as a significant producer of) fossil fuels. As a result, it is clear that the Tribe was not resisting only one project but was forced to contend with broader trends relating to the U.S. economy, entrenched fossil fuel consumption, and a history soaked in oil. It is for this reason that the Standing Rock fight spurred more conversations about renewable energy in the U.S., for many saw the irony inherent in driving to the protest site, or sending packages by air, to support an effort to *stop* oil from flowing. This is not to say that the Tribe could not win their fight through both public opinion and court battles (the latter of which, as of this writing, remains to be seen). Indeed, Klein (2014) and LaDuke (2017a) recognize the growing power of Indigenous resistance against fossil fuels in Canada and the U.S. However, it is important to recognize that because we as a nation are "addicted to oil," the Tribe also was going up against a way of life almost wholly predicated on the consumption of fossil fuel resources.

Scope of the book

This book explores the interconnection of media, environmental justice, U.S. politics, and heightened global concerns about environmental issues through a particularly fruitful and provocative lens: the Standing Rock struggle over the Dakota Access Pipeline (commonly known as the "#NDAPL" or "#NODAPL" movement). The focus of this work stays consistently on how the news media covered the movement, with one of the first objectives being to outline some of the central issues in commercial journalism in the U.S., a discussion built around what I am terming "deep media." Working from this foundation, this book then provides a contemporary evaluation, through a political economy perspective, of how the news media system has covered environmental justice issues. This enables not only a useful scholarly update in this area but also lays the groundwork to evaluate how journalism's drive for profit, especially in the U.S., impacts coverage of the environment (broadly) and environmental justice issues (in particular).

The central focus of the book can be divided into two parts, with the first being a framing analysis conducted on coverage of #NODAPL by news organizations in the U.S. and Canada. Frames reflect "the play of power and boundaries of discourse over an issue" (Entman, 1993, p. 55), which are especially important to reveal when it comes to issues concerning environmental justice. Essentially, the framing analysis provides a point of comparison between various outlets within each country and then also a broader juxtaposition between the two countries, both of which have their own distinct relationship to fracking, environmental issues, and Indigenous nations. Centrally, the analysis also reveals how the Indigenous "water protectors," local law enforcement, and the pipeline itself were represented by the news media. While media coverage of Indigenous environmental justice movements matters for many reasons, one of the most important is that coverage can draw attention to pressing political, social, and environmental issues. As Entman (1993) observes, one of the most powerful "frames" in journalism is that of omission: without sufficient attention by the news media, any issue, regardless of its importance, can languish in obscurity.

The second key research component is comprised of interviews with members of the Standing Rock Sioux Tribe. In these interviews, tribal members, including the current chairman, reveal how they feel about the pipeline, their history, the U.S. government (including President Trump), Big Oil, and where they are headed as a nation. I also interviewed Mark Trahant, a journalist, editor, and member of Idaho's Shoshone-Bannock Tribe who has four decades of national and international experience. Together, these two lines of research (framing analysis and interviews) provide powerful, complementary, and (at times) surprising views on news media coverage of environmental justice issues. The discussion then expands to considering the findings in light of current U.S. politics, including a Trump presidency that employs "law and order" rhetoric regarding people of color and that often subjects environmental issues to an economic "cost-benefit" analysis. The book concludes by briefly considering the role of social media in the era of "Big Oil" and growing Indigenous resistance and power.

Background on the Dakota Access Pipeline and the Standing Rock resistance

The Dakota Access Pipeline was first conceived of in 2014 as a way for Energy Transfer Partners (ETP), a Texas-based fossil fuel infrastructure company on the

Fortune 500 list, to move fracked oil from the expansive North Dakota Bakken fields to Illinois through a 1,172-mile pipeline built by ETP's subsidiary Dakota Access LLC. From the beginning the pipeline has been controversial due to issues including individual homeowners' concerns about the pipeline on their property and the early environmental review process that critics claimed did not consider climate change and significant oil spills. Aisch and Lai (2017) at the *New York Times* provide perhaps the best visual treatment of controversies along the length of the pipeline, including a lawsuit by the Yankton Sioux tribe against the U.S. Army Corps of Engineers and the contention that the pipeline crosses near (or over) land originally ceded to the Standing Rock Sioux Tribe in the 1851 Treaty of Fort Laramie. Additional disputes include a lawsuit from nine landowners in Iowa who sued the Iowa Utilities Board and Dakota Access LLC to protect their land from the government's claim of eminent domain (Petroski, 2016).

Resistance to the Dakota Access Pipeline

Undoubtedly the most well-known controversy to date is the potential negative impact on the water supply for the Standing Rock Sioux Tribe. The resistance to the pipeline was initiated in April of 2016 by a youth group called the "One Mind Youth Movement" founded by Lakota Sioux youth Jasilyn Charger as well as Trenton Casillas-Bakeberg and Joseph White Eyes (Elbein, 2017). The group created a small "prayer camp" on the northern side of the Standing Rock Sioux Reservation in North Dakota. In conjunction with this, the youth members also planned two youth runs (a 500-mile run from Sacred Stone Camp to Omaha and a much longer one from Sacred Stone to Washington, DC) with the ultimate goal of delivering a message to the U.S. Army Corps of Engineers: stop the pipeline project. By then their run had attracted the attention of other organizations like the Indigenous Environmental Network, and by August, a larger number of people began arriving at the prayer camp.

That same month, the Standing Rock Sioux Tribe sued the Army Corps for violating the National Historic Preservation Act and for not adequately consulting with the Tribe before approving the project (Hersher, 2017). By September, the pipeline resistance was well underway and journalist Amy Goodman with *Democracy Now!* published the video on social media showing security personnel for Energy Transfer Partners using dogs and mace on the demonstrators. Her coverage seemed to mark a turning point in the way mainstream media outlets covered the demonstrations, and more news organizations ended their media blackout and began to visit Standing Rock.

In early December 2016, as the protests were in full swing, the Army Corps of Engineers, under President Obama, temporarily halted the project by denying Dakota Access LLC the easement needed to excavate under the Missouri river. However, two months later in early February 2017, recently inaugurated President Trump ordered the Army Corps to issue the final permit needed for ETP to complete the pipeline. The story did not – and does not – end there, for in June of 2017 a federal judge ruled that the U.S. government conduct a more lengthy environmental review process, writing that the Army Corps "did not adequately consider the impacts of an oil spill on fishing rights, hunting rights, or environmental justice, or the degree to which the pipeline's effects are likely to be highly controversial" (Hsu, 2017). As of this writing, the environmental review is ongoing.

Environmental justice, the Standing Rock Sioux Tribe, and human rights

It is important to recognize at the outset that the protests of the Standing Rock Sioux and other Tribes over the Dakota Access Pipeline are important in their own right. That is, issues concerning Indigenous groups' access to clean water and land rights can be considered a standalone issue worthy of exploration that could be placed within a broader pattern of environmental justice cases in the U.S. (see especially Grossman & LaDuke, 2017; Moore & Lanthorn, 2017; Pasternak, 2010; Adamson, 2001; Weaver, 1996) and Canada (e.g., LaDuke, 2017b; Wiebe, 2016; Klein, 2014; Agyeman et al., 2009). However, it also is the case that the Tribe's struggle is emblematic of broader issues that invite additional examination.

First, the events surrounding #NODAPL have reinvigorated a nationwide discussion about renewable energy and the environmental dangers of continued fossil fuel use, especially in regard to climate change and water pollution. Second, the pipeline controversy also has underscored the urgent need to talk about the interwoven concerns of racial tension, human rights, and environmental justice. In November 2016, the *U.S. Commission on Civil Rights* issued the following statement expressing its dismay about the events unfolding at Standing Rock:

> We are concerned with numerous reports and testimony regarding the use of military-style equipment and excessive force against protesters. Protesters have a constitutional right to peacefully assemble and lawfully express their concerns about the environmental and cultural impacts of the pipeline. Our concerns are compounded by the disproportionate police use of excessive force against Native Americans, who are more likely than any other racial group to be killed by police. We call upon federal, state, and local officials and law enforcement to work together to de-escalate the situation and guarantee the safety of protesters to exercise their First Amendment rights.[3]

Many of the images that have come out of the pipeline protests, including dogs biting Indigenous people and water cannons being used against protesters, are evocative of the intense civil rights struggle in Birmingham, Alabama in 1963. As Rebecca Solnit (2016) with *The Guardian* has observed, "What's happening at Standing Rock feels like a new civil rights movement that takes place at the confluence of environmental and human rights and grows from the last 60 years of lived experience in popular power and changing the world." I believe Solnit is correct in this, but it also is true that the manifestation and character of this larger movement is yet to be fully understood. As I write this, a woman protesting a white supremacist march in Charlottesville, Virginia, where many of the marchers were yelling racial slurs, was killed by a white nationalist who drove his car into a crowd. Confederate statues are being removed by force (or, alternately, shrouded) as symbols of a painful, racist past marked by unequal power. Of course, all of this has occurred alongside the political upheaval created by a majority-white populist movement that swept Donald Trump into power.

This research thus comes at a time of deep, emotional discussions in the U.S. about the rights of people of color, including not only Indigenous groups but also members of African-American and migrant communities. This book enters this heady mixture of political and socio-cultural upheaval in the U.S. to examine the role of the media in an environmental justice case like Standing Rock: what role does a commercial media system (which is in flux itself) play in these types of conflicts? How are Indigenous groups like the Standing Rock Sioux Tribe represented in the news media? How does the emerging role for social media in social movements influence the information the public receives about important issues relating to environmental justice? And, perhaps most importantly: how does the Standing Rock Sioux Tribe perceive of the media coverage during the resistance against the Dakota Access Pipeline?

Qualitative inquiry: the critical interpretive framework of this research

The qualitative research that comprises the foundation of the research in this book is inductive, applied, deeply critical, and engaged – four elements that are inextricably and intentionally intertwined here. In addition, this work draws from different qualitative methods (including interviews and framing) and many different approaches within the broader field of communication research. As Denzin and Lincoln (2018, p. 12) remind us, "qualitative research, as a set of interpretive activities, privileges no single methodological practice over another [and] has no theory or paradigm that is distinctly its own." In the same vein, Kovach (2018) observes that what may be considered "Indigenous research" spans many disciplines and can involve lines of inquiry that employ numerous different approaches: "Indigenous research concerns itself with Indigenous matters, although it may or may not involve itself with Indigenous peoples… Indigenous research can be viewed as an umbrella term that includes myriad research opportunities" (p. 215). In this sense, the current research can be considered broadly as Indigenous research in its interdisciplinarity as well as its focus on how the media covered the Standing Rock Sioux Tribe's struggle for environmental justice.

The critique undergirding this research is rooted in the principles of political economy, which, in its broadest sense, examines and explores the economics undergirding the media industry. Meehan et al. (1993) in particular identify a focus on the corporate structure, ownership, financing, and "market structure" of the media – to understand their influence on technology as well as politics and culture. In addition, this approach could be more specifically identified as *critical political economy*, which Hardy (2014, p. 3) defines as "a tradition of analysis that is concerned with how communication arrangements relate to goals of social justice and emancipation." This critical perspective still examines the ownership, production, and distribution of the media system but does not see the economic sphere as separate from cultural, social, and political phenomena. Instead, it steers us towards the idea of social justice and "emancipation," focuses on the "unequal distribution of power," and includes serious consideration of the commodification and industrialization of culture (Hardy, 2014, p. 6). This approach is especially relevant in the consideration of how economics in a broad sense impacted journalism's coverage of the #NODAPL movement and of the Standing Rock Sioux Tribe itself. Which factors of the media industry are at play, and how did this influence the way in which the Tribe and the movement were depicted by the press? As Meehan et al. (1993, p. 105) argue, "in the U.S., where media are always and primarily businesses, the (political economy) approach has been crucial to understanding why we get what we get" from the media industry.

I chose this research because the Tribe's struggle had captured my attention, and not simply for academic reasons relating to my identity and interests as a media scholar. Instead, the movement at Standing Rock connected with my ongoing support of human rights issues and environmental justice. However, as is often the case, real-world concerns align with academic efforts, and thus my graduate students and I at the University of Washington Tacoma changed our research focus in the autumn of 2016 to examine what was happening at Standing Rock. Our resulting academic endeavors then turned into research presentations at an international conference in the U.K.,[4] and the idea for this book was born. Although this research falls under the category of academic scholarship, it is my hope that those who similarly found themselves interested in – and engaged with – the #NODAPL movement will read this book as well. This desire is in keeping with what Wasko (2014, p. 268) has observed about political economy of media (PEM) researchers, in that "many (if not most) PEM scholars incorporate issues related to policy and activism in their research, as well as working outside academic settings to promote media change, as well as social change generally."

Along these lines, this research can be considered a part of what is termed "applied communication" or "applied research." A central idea undergirding "applied research" is that the problems it seeks to understand are grounded in the "real world" instead of remaining abstract. Lindlof and Taylor (2019) write about this cogently, observing that those who do applied research are "motivated to adapt and innovate qualitative methods that empower underserved groups" (p. 28). As Michael Lawson (2009) writes in *Dammed Indians Revisited* (about the impact of the Pick–Sloan Plan to create a series of dams along the Missouri River), his work was "the greatest fulfillment of my career that what began as a purely academic exercise for me... eventually found a real-world application...that significantly benefited the Sioux people" (p. xxiii). Although I don't believe I can compare my current analysis of media coverage of the #NODAPL movement to Lawson's decades-long, active, hands-on role[5] regarding the "involuntary resettlement" of many Indigenous groups, including members of the Great Sioux Nation, my hope for this book is to address what Leistyna et al. (2005) term the *representational assault* by the media[6] on not only the "water protectors" at Standing Rock but also Indigenous people more broadly. Dempsey et al. (2011, p. 258) argue that:

> A basic assumption of social justice scholarship is the need for transformative social change.... Social justice scholarship brings with it a diagnosis and a call to action. Through the analysis of oppression, social justice scholarship works to reclaim suppressed conflicts, opening them up to scrutiny. However, scholars take more or less activist orientations in pursuing the commitment to dismantling systems of inequality.

From this perspective, then, this research has action and engagement at its center: it is oriented towards social justice in a way that aligns with another feature of this qualitative research in that it operates within a critical theoretical framework. Lindlof and Taylor (2019, p. 70) provide an admirably thorough list of the premises inherent in critically oriented scholarship, noting that critical scholars perceive theory and practice as inextricably connected; define research as inherently political "in their potential for affecting power relations in society"; reject positivist approaches that presume objectivity; "serv[e] cultural groups by fostering their recognition of... politics that shape their situations"; engage in critical reflection of their own roles and motivations in research; and challenge neoliberal conceptions of democracy. In addition, this work attempts to heed the warning by Hopi scholar and educator Lomayumtewa Ishii (2011) to avoid *intellectual colonialism*, whereby "meaning and authority are constructed and maintained within the colonizer's epistemological...activities, including research, publishing, and intellectual discursive practices and formations, and in controlling the acceptability of cultural referents for popular intellectual consumption" (p. 54).

Outline of chapters

Chapter Two ("The Dakota Access Pipeline, the oilygarchy, and the media: a study in power") begins to answer the questions outlined in this first chapter through a broad examination of the commercial journalism landscape through a critical political economy perspective. In particular it traces the economic and political contours of the U.S. media landscape as a way to discuss its economic motivations as well as its flaws and vulnerabilities. It is in this chapter that the concept of *deep media* is defined and brought to bear on a controversial event like the Standing Rock protests. *Deep media*, as envisioned here, places less emphasis on the "biases" or motivations of individual players (journalists, editors, media owners) in the media industry while stressing the need to understand media coverage of #NODAPL through the lens of structural deficiencies in the media landscape itself. This concept then provides a foundation for the remainder of the book, culminating in the final chapter where the ability of social media to address some of the structural weaknesses in commercial news media is discussed.

Chapter Three ("A case of un-coverage? *Deep media*, Indigenous representation, and environmental issues") builds on the political economy critiques in the previous chapter to explore how a commercial, largely corporate-owned U.S. media system covers a topic like the environment (in general) and environmental justice (in particular). The premise of this chapter builds from previous work recognizing that the environment is covered infrequently, episodically, and with a focus on drama (Moore & Lanthorn, 2017; Boykoff & Boykoff, 2007; Nambiar, 2014; Beder, 1998; Anderson, 1997). Included here is a brief discussion of which environmental issues are most prominent today and which ones are most often covered by the mainstream press, in the process identifying key tensions between commercialism, journalism, and coverage of the environment. The chapter surveys key examples of the relationship between journalism and environmental justice from around the globe (including Jill Hopke's work on El Salvadorean media, Tran Thi Thuy Binh's documentation of metal refineries in Vietnam, and Ghanaian e-waste journalist Mike Anane on the infamous Agbogbloshie e-waste site in Western Africa) as a way to contextualize media coverage of the events at Standing Rock. Here, the case is made for the #NODAPL movement being both unique in its struggle and media coverage, and similar to Indigenous struggles over environmental injustice occurring elsewhere in the world.

Chapter Four ("Framing injustice: U.S media coverage of the Standing Rock movement") presents the results of framing analysis of #NODAPL by news media outlets from the U.S. In this chapter, framing is clearly defined as both a method and a theoretical framework, and the parameters of the research are described fully. Then, a framing analysis is conducted on three news sources: the *New York Times*, *Indian Country Media Network*, and the *Bismarck Tribune*. From this emerges a discussion of the *Law and Order* frame as used by local media outlets in particular as well as the unexpected shift in framing of #NODAPL from the *New York Times* that occurred after Amy Goodman's viral video on social media.

Chapter five ("'Could it happen here?' Canadian newspaper framing of the Dakota Access Pipeline") places the focus on Canadian newspapers by identifying the key frames visible in several provincial and national newspapers (the *Globe and Mail*, *National Post*, and *Calgary Herald*). This assessment is meant to expand the analysis to include an international perspective and as a way to compare results given each country's history and experience with fracking, Indigenous people, and fossil fuels. While it is true that the Canadian press differed slightly from the U.S. in terms of the *types* of frames themselves, it also is the case that there was a *quantitative* difference, as it was difficult to find Canadian dailies that covered the #NODAPL movement at all. As noted earlier, framing research considers what is omitted from coverage just as important as what is included, and thus the relative absence of press attention – especially for a country heavily invested in fracking, with numerous First Nations populations, and with a border close to Standing Rock – is considered a significant finding in itself.

Chapter Six ("Law and order: from civil rights to Nixon to Trump, a trope in revival") connects the specific frame of *Law and Order* found in U.S. and Canadian newspapers to American history and contemporary politics. The chapter includes a discussion of media coverage and political upheaval during the civil rights era and "The Long Hot Summer" in the U.S. as a way to understand how U.S. politicians described those demonstrations as a breakdown of order in society. Undergirding this discussion is the recognition of former U.S. President Nixon as the original "Law and Order" president when it came to conflict, civil disobedience, and people of color in the U.S. This discussion then lays the foundation to understand where we are now with President Trump's frequent employment of the "law and order" approach and support for militarized police actions such as those seen in North Dakota. The chapter ends with a discussion of the implications of a "law and order" revival by politicians and the news media for the struggle over the Dakota Access Pipeline.

In Chapter Seven ("Indigenous perspectives on the Dakota Access Pipeline, politics, and the media: the Standing Rock Sioux Tribe and journalists speak"), interviews with the Standing Rock Sioux Tribe reveal their unique experiences and observations on media coverage of their protests. While the outside observer may see resistance to the Dakota Access Pipeline as a standalone event, the members of the Tribe with whom I spoke only saw the pipeline through the long lens of history. As such, they tied the events surrounding the pipeline to issues of sovereignty, to renewed solidarity with other Indigenous groups worldwide, to their lengthy and complex history with the U.S. government, and to their future. In addition, the voice of experienced journalist and scholar Mark Trahant (a member of Idaho's Shoshone-Bannock Tribe) is included here. A professor at the University of North Dakota School of Communication when I spoke with him (he has since accepted the role of editor at *Indian Country Today*), Mr. Trahant provided his broad perspective on the press, his more specific views on the coverage of #NODAPL and oil pipelines, and even his experience interviewing past U.S. presidents and members of their administrations. Because Indigenous voices are rarely covered in mainstream media – by both journalism and popular culture – this is a valuable opportunity to gain new perspectives and reveal points of intersection between my own media analysis and their firsthand observations.

Chapter Eight ("Did technology kill the goose that laid the golden egg or save it? New media, old media, and the #NODAPL movement") places primary focus on exploring how new media and the Indigenous press can challenge traditional journalism to pay attention to environmental justice issues and create important firsthand *testimony* that can shape public perceptions and action. The chapter begins with a return to the viral video by *Democracy Now!* of the dogs biting the pipeline demonstrators, exploring the implications of social media in the context of traditional journalism. One of the central questions addressed in this section revolves around the question of whether the old media–new media relationship is symbiotic or adversarial. As such, this chapter examines the shifting terrain of the media landscape as a way to understand where we find ourselves now through the lens of the Standing Rock movement. Perhaps instead of seeing that technology "killed the goose that laid the golden egg," as several scholars have noted in the past in regard to journalism, social media represent a potentially liberatory and democratizing phenomenon.

Regarding the order of chapters

Each of these chapters is meant to build on the previous ones, moving from a broad understanding of the economic and political foundations of the U.S. press and how it covers environmental issues to a study of how the media covered the protests over environmental justice at Standing Rock. Having noted this, it also is the case that those less interested in the political economy of media – and perhaps more interested in learning about the perspectives of the Standing Rock Sioux Tribe or the #NODAPL movement – can visit those chapters first. For those who prefer to learn more about the interplay between social media and social movements, the last chapter may hold more interest. As such, the book can be read in a variety of different ways depending on the reader's interest and inclination, and I do not consider there to be any one "right" way to progress through this work.

Notes

1 Garrett and Weeks (2017) provide a thorough evaluation of the tense relationship between feelings, facts, and political beliefs in the era of "fake news."
2 Cusick cites a 2013 report from the *Worldwatch Institute*, citing Gonzalez and Lucky (2013).
3 Titled "The U.S. Commission on Civil Rights concerned with Dakota Access Pipeline," the full document is available at http://www.usccr.gov/press/2016/PR-11-22-16-Dakota-Pipeline.pdf
4 Their work included "Fighting the Black Snake: Big Oil, censorship, and democracy" by Rachael Williamson; "The Dakota Access Pipeline: Why is access to clean water still up for negotiation? by Katherine Jennison; and "'Bad' Indians versus 'good' business: Media colonialism in news coverage of the Dakota Access Pipeline protest" by Erica Tucker.
5 Lawson drafted historical reports, submitted reports to the government, and spoke at Congressional hearings that resulted in historic legislation, funds, and settlements for several Tribes.
6 Hall (1997) terms this the "politics of representation," which recognizes the significant power of media representations of people of color.

References

Adamson, J. (2001). *American Indian literature, environmental justice, and ecocriticism: The middle place*. Tucson, AZ: University of Arizona Press.

Agyeman, J., Cole, P., Haluza-DeLay, R., & O'Riley, P. (2009). *Speaking for ourselves: Environmental justice in Canada*. Vancouver, BC: UBC Press.

Aisch, G., & Lai, R. (2017, March 20). The conflicts along 1,172 miles of the Dakota Access Pipeline. *New York Times*. Retrieved from https://www.nytimes.com/inter active/2016/11/23/us/dakota-access-pipeline-protest-map.html?mcubz=3

Anderson, A. (1997). *Media, culture and the environment*. London: Routledge.

Beder, S. (1998). *Global spin: The corporate assault on environmentalism*. White River Junction, VT: Chelsea Green Pub.

Boykoff, M. T., & Boykoff, J. M. (2007). Climate change and journalistic norms: A case-study of US mass-media coverage. *Geoforum*, *38*(6), 1190–1204. 10.1016/j.geoforum.2007.01.008

Bullard, R. (2000). *Dumping in Dixie: Race, class, and environmental quality* (3rd ed.). Boulder, CO: Westview Press.

Cusick, D. (2013, October 25). Fossil fuel usage continues to rise. *Scientific American*. Retrieved from https://www.scientificamerican.com/article/fossil-fuel-use-contin ues-to-rise/

Dalrymple, A. (2016, August 18). Pipeline route plan first called for crossing north of Bismarck. *Bismarck Tribune*. Retrieved from https://bismarcktribune.com/news/st ate-and-regional/pipeline-route-plan-first-called-for-crossing-north-of-bismarck/ article_64d053e4-8a1a-5198-a1dd-498d386c933c.html

Dempsey, S., Dutta, M., Frey, L. R., Goodall, H. L., Madison, D. S., Mercieca, J., & Miller, K. (2011). What is the role of the communication discipline in social justice, community engagement, and public scholarship? A visit to the CM café. *Communication Monographs*, *78*(2), 256–271. 10.1080/03637751.2011.565062

Denzin, N., & Lincoln, Y. (2018). Introduction. In N. Denzin & Y. Lincoln (Eds.), *The SAGE handbook of qualitative research* (5th ed.). Thousand Oaks, CA: SAGE.

Elbein, S. (2017, Jan 31). The youth group that launched a movement at Standing Rock. *New York Times*. Retrieved from https://www.nytimes.com/2017/01/31/magazine/ the-youth-group-that-launched-a-movement-at-standing-rock.html

Entman, R. M. (1993). Framing: Toward clarification of a fractured paradigm. *Journal of Communication*, *43*(4), 51–58. 10.1111/j.1460-2466.1993.tb01304.x

Garrett, R., & Weeks, B. (2017). Epistemic beliefs' role in promoting misperceptions and conspiracist ideation. *PLoS One*, *12*(9), e0184733. https://doi.org/10.1371/journal. pone.0184733

Grossman, Z., & LaDuke, W. (2017). *Unlikely alliances: Native nations and white communities join to defend rural lands*. Seattle, WA: University of Washington Press.

Hall, S. (1997). *The spectacle of the "Other"*. London: Sage.

Hardy, J. (2014). *Critical political economy of the media: An introduction*. London: Routledge.

Hersher, R. (2017, February 22). Key moments in the Dakota Access Pipeline fight. National Public Radio. Retrieved from https://www.npr.org/sections/thetwo-way/ 2017/02/22/514988040/key-moments-in-the-dakota-access-pipeline-fight

Hsu, S. (2017, June 14). Federal judge orders environmental review of Dakota Access Pipeline. *Washington Post*. Retrieved from https://www.washingtonpost.com/local/ public-safety/federal-judge-orders-environmental-review-of-dakota-access-pipel ine/2017/06/14/6de94c98-5152-11e7-b064-828ba60fbb98_story.html?utm_term=. 15cfd069c2c0

Ishii, L. (2011). Hopi culture and a matter of representation. In S. Miller & J. Riding (Eds.), *Native historians write back: Decolonizing American Indian history* (pp. 52–68). Lubbock, TX: Texas Tech University Press.

Jackson, W. (2010). *Consulting the genius of the place: An ecological approach to a new agriculture*. Berkeley, CA: Counterpoint Press.

Juhasz, A. (2008). *The tyranny of oil: The world's most powerful industry—and what we must do to stop it*. New York: William Morrow.

Klein, N. (2014). *This changes everything: Capitalism vs. the climate*. New York: Simon & Schuster.

Kovach, M. (2018). Doing Indigenous methodologies. In N. Denzin and Y. Lincoln (Eds.), *The Sage Qualitative Handbook of Qualitative Research* (5th ed., pp. 214–234). Thousand Oaks, CA: SAGE.

LaDuke, W. (2017a, January 31). Winona LaDuke on new ways to keep pipelines out of the great lakes. *Yes! Magazine*. Retrieved from http://www.yesmagazine.org/planet/tribes-find-new-ways-to-keep-pipelines-and-their-oil-out-of-the-great-lakes-20170131

LaDuke, W. (2017b). *The Winona LaDuke chronicles: Stories from the front lines in the battle for environmental justice*. Nova Scotia: Fernwood Publishing.

Lawson, M. L. (2009). *Dammed Indians revisited: The continuing history of the Pick-Sloan Plan and the Missouri River Sioux*. Pierre, SD: South Dakota State Historical Society Press.

Leistyna, P., Alper, L., & Asner, E. (2005). *Class dismissed: How TV frames the working class*. [Video/DVD] Northampton, MA: Media Education Foundation.

Lindlof, T. R. & Taylor, B. C. (2019). *Qualitative communication research methods* (4th ed.). Thousand Oaks, CA: SAGE Publications, Inc.

Meehan, E., Mosco, V., & Wasko, J. (1993). Rethinking political economy: change and continuity. *Journal of Communication, 43*(4), 105–116.

Moore, E. E., & Lanthorn, K. R. (2017). Framing disaster. *Journal of Communication Inquiry, 41*(3), 227–249. 10.1177/0196859917706348

Morris, I. (2015). *Foragers, farmers, and fossil fuels: How human values evolve*. Princeton, NJ: Princeton University Press.

Nambiar, P. (2014). *Media construction of environment and sustainability in India*. Los Angeles, CA: SAGE.

Newman, P., Beatley, T., & Boyer, H. (2017). *Resilient cities: Overcoming fossil fuel dependence* (2nd ed.). Washington, DC: Island Press.

Oremus, W. (2016, December 6). Stop calling everything "fake news." *Slate Magazine*. Retrieved from http://www.slate.com/articles/technology/technology/2016/12/stop_calling_everything_fake_news.html

Pasternak, J. (2010). *Yellow dirt: An American story of a poisoned land and a people betrayed*. New York: Free Press.

Petroski, W. (2016, April 11). Iowa landowners file suit over Bakken pipeline. *Des Moines Register*. Retrieved from http://www.desmoinesregister.com/story/news/politics/2016/04/11/iowa-landowners-file-suit-over-bakken-pipeline/82915590/

Sampson, A. (1975). *The seven sisters: The great oil companies and the world they shaped*. New York: Viking Press.

Solnit, R. (2016, September 12). Standing Rock protests: This is only the beginning. *The Guardian*. Retrieved from https://www.theguardian.com/us-news/2016/sep/12/north-dakota-standing-rock-protests-civil-rights

Thomas, Z. (2016, January 22). The global oil glut is squeezing the US shale industry. *British Broadcasting Corporation*. Retrieved from http://www.bbc.com/news/business-35355286

Thorbecke, C. (2016, November 3). Why a previously proposed route for the Dakota Access Pipeline was rejected. *American Broadcasting Company.* Retrieved from https://abcnews.go.com/US/previously-proposed-route-dakota-access-pipeline-rejected/story?id=43274356

Wasko, J. (2014). The study of the political economy of the media in the twenty-first century. *International Journal of Media & Cultural Politics, 10*(3), 251–279.

Weaver, J. (1996). *Defending mother earth: Native American perspectives on environmental justice.* Maryknoll, NY: Orbis Books.

Wiebe, S. M. (2016). *Everyday exposure: Indigenous mobilization and environmental justice in Canada's chemical valley.* Vancouver, BC: UBC Press.

2

THE DAKOTA ACCESS PIPELINE, THE OILYGARCHY, AND THE MEDIA

A study in power

The preceding chapter posed questions regarding the role of mainstream news media in the depiction of Indigenous environmental justice struggles – questions that form the focal point of this book. Before beginning to answer those questions, it is important to trace the interrelated contours of power in the pipeline conflict. As even the brief timeline of developments makes clear, the Dakota Access Pipeline is steeped in power struggles, ones that emerge in sharp relief from the change in political power that occurred during the 2016 U.S. presidential election. The issues, of course, run deeper than just electoral politics, for these struggles are enmeshed in the powerful web created by the oil industry, the U.S. government, and the professional news media system. In concert with the Standing Rock Sioux Tribe and Energy Transfer Partners, these form the primary players in the pipeline struggle.

This chapter thus draws significant lines of connection between the oil industry, the U.S. government, and the commercial news media system as foregrounding to understand the Indigenous resistance that formed around the Dakota Access Pipeline. As becomes clear, corporate interest (especially that relating to the oil industry) rests at the heart of not only the #NODAPL movement but also the government and professional news media systems. These interconnections likely will be evident to serious scholars of the media, especially those well versed in political economy; however, to make the relationships explicit here is to begin to effectively contextualize the Standing Rock Sioux environmental justice struggle.

Big Oil: an old titan with political and environmental impact

In 1928, the heads of three oil companies – Royal Dutch Shell, British Petroleum, and Exxon – met at Achnacarry Castle in Scotland for a few days of grouse hunting. Unbeknownst to the public, they were also quietly creating the beginnings of a powerful oil cartel whose goal was to set and maintain market shares (Gareau, 2002). Eventually, four other oil companies joined them in what became known as the "Seven Sisters,"[1,2] a cartel of oil companies run by wealthy Americans and Europeans that agreed to collude instead of to compete with each other (Sampson, 1975; Yergin, 2008). The result became known as the "Gulf Plus" pricing structure, which was significant because

> it set the price of crude, no matter where purchased, at the price of crude existing in the United States plus the standard charge of shipping the product from the Gulf of Mexico to the particular market in question. The effect of this bizarre system was to stabilize oil prices and to allow the oil companies to reap extraordinary profits.
>
> *(Gareau, 2002, p. 82)*

Dominating the oil industry meant that the "Sisters" wielded influence over "virtually every country in the world" (Gareau, 2002, p. 83). As Sampson (1975) notes, "with their complexity, range, and resources, they were institutions that had appeared to be a part of world government" (p. 5). The result laid the foundation for the power structure of "Big Oil" today, for even though the Sisters' power eventually waned, they had forged a pathway for a new and powerful global oil industry that occurred through a series of giant mergers.

Juhasz (2008) reveals the contemporary power of the fossil fuel industry in her book *The Tyranny of Oil*,[3] noting that mergers (all of which have occurred since 1999) have

> helped Big Oil establish its footing as a major owner of oil. While nowhere near its Seven Sisters 'glory years,' Big Oil's oil reserves are impressive... Were the five largest oil companies operating in the United States one country instead of five corporations, their combined crude oil holdings would today rank within the top ten of the world's largest oil-rich nations.
>
> *(p. 6)*

What is so oppressive about big oil, Juhasz contends, is that it creates significant "environmental pollution, public health risks, and climate destruction...at every stage of oil use, from exploration to production, from transportation to refining, from consumption to disposal" (p. 2). Along each stage, Juhasz contends, the "masters of the oil industry" exert influence on politics, democracy, war, and the environment[4]:

> Big Oil has simple needs. It wants to explore for, produce, refine, and sell oil and gas wherever possible without restriction. It wants laws that allow it to expand all of its operations. It wants to prevent laws that stand in its way and roll back those that already exist. Big Oil wants friends in office and enemies out of office. It wants friendly regulators in government...It

does not want to be slowed down or financially burdened by government bureaucracy, environmental laws… or concerns for human rights.

(pp. 209–210)

From tax loopholes to key legislation, the oil industry has enjoyed a "powerful sway over American politics," one made manifest in the organized, systematic, corporate-fueled climate change denials in the last few decades (Mayer, 2016, p. 249).

As a result, there are few arguments today denying the immense power and influence of what journalist Amy Goodman (2017) is fond of calling the "oilygarchy."[5] *Oilygarchy* is, of course, a clever transformation of the word "oligarchy," which refers to power held in the hands of a few.[6] Fossil fuels have permeated our lives in the U.S. to such a degree that Princen et al. (2015, p. 7) note that "keeping fossil fuels in the ground" it is often considered "unthinkable" in our society.[7] It is the significant lines of connection between the oil industry and the U.S. government (especially the notion of the "deep state") where we turn our attention to next, for revealing these latent yet powerful relationships is at the heart of understanding the struggle by the Standing Rock Sioux Tribe against the oil pipeline.

The oilygarchy and the "deep state"

Numerous scholars over many years have described what has become known as the "deep state":[8] namely, that while there is a visible, government that citizens recognize and feel that they know, there then is a more obscured version of government that contains its own structures and logic, one where the real power lies and where the real decisions are made (Lofgren, 2016; Scott, 2015; Engelhardt, 2014; Ambinder & Grady, 2013). One of the key elements of what is also referred to as a "shadow government" is the gatekeeping function, where key information is kept from the public, ostensibly for public safety and the efficient running of government (Ambinder & Grady, 2013). Lofgren (2016) describes *deep state* as "the subsurface part of the iceberg [that] operates on its own compass heading regardless of who is formally in power" (p. 30). He is quick to point out that he considers the deep state more akin to an evolution of a large bureaucracy than a conspiracy. And while he believes deep state to be a "hybrid of national security and law enforcement agencies, plus key parts of the other branches whose roles give them membership" (p. 34), he gives final credit to corporate influence, observing that: "it is not too much to say that Wall Street may be the ultimate owner of the Deep State and its strategies" (p. 36).

In *The American Deep State: Wall Street, Big Oil, and the Attack on U.S. Democracy*, Peter Scott (2015) seems to agree, arguing that in the U.S., "our society, by its very economic success and consequent expansion, has been breeding impersonal forces both outside and within itself that are changing it from a bottom-up elective democracy into a top-down empire" (p. 2). In Scott's discussion of what he calls a "two-level" state, he cites Connecticut Attorney General Robert Killian's statement that "I hate to say that Big Oil is bigger than the United States Government, but its favored treatment at the hands of our government certainly leads to that conclusion" (p. 11).

Scott believes the deep state is the result of an "ambiguous symbiosis" between U.S. national security agencies (like NSA and CIA) and "the much older power of Wall Street," including the law firms and powerful banks within that circle (2015, p. 13).[9] Returning to the oilygarchy, Scott identifies the power of oil cartels in several examples, including when President Truman in 1952 tried (and failed) to break up the Seven Sisters that were controlling the distribution of global oil. The Sisters were denying Iran the right to nationalize its own oil supplies from the British.[10] After manufacturing a successful boycott of Iranian oil, the Sisters then set about convincing the CIA to overthrow the Iranian Premier Mossadegh, who had been instated by democratic election, which then led to decades of instability in the country (Berman, 2006). Scott thus sees "the oil cartel as a structural component of the American deep state" (p. 20). Even when the Sisters' power and influence had waned, Scott notes, the power of large oil interests continued to grow, especially in regard to the high gas prices in the U.S. between 1979 and 1980. While numerous scholars have attributed the oil and gas scarcity to "political upheavals in Iran," he notes that a more likely explanation is that "American oil companies, not Iranian turmoil, were primarily responsible for the gas shortages" (p. 28).

What becomes clear from discussions about a "deep state" is that its origins lie less in an outright conspiracy or "evil" coterie of political elites and more in the structural deficiencies in government – especially those that render the U.S. government more vulnerable to corporate influence. Thus, although no scholar seems to define "deep state" exactly the same way, the idea of corporate control (especially from the oil industry) of government is a common thread. Other examples regarding the relationship between government and the oil industry abound. Shrivastava and Stefanick (2015) note that reliance on oil revenues by the provincial government of Alberta created a "symbiotic relationship" between the oil industry and the Canadian government[11] – a situation that has clear parallels to North Dakota and its lucrative, oil-rich Bakken shale. In addition, although big monied interests like oil industry colossus ExxonMobil have a stake in the Dakota Access Pipeline, much of the power behind the pipeline came from several financial industry giants, including Citibank and Wells Fargo in the U.S., TD Bank in Canada, and Mizuho bank from Japan (Tabuchi, 2016). Because of the financial muscle behind the pipeline, #NODAPL protesters went after them. In one prominent example, the city of Seattle, Washington responded to the protests by divesting $3 billion from Wells Fargo over the pipeline construction's impact on human rights. Tabuchi (2016) writes,

> In campaigning to reduce the world's carbon emissions, environmentalists have increasingly focused on the financiers behind the fossil fuel industry –

highlighting their role in financing coal, oil and gas projects. It is an expansion of traditional protest efforts, and it has met with some early success.

(Tabuchi, 2016)

Recognizing the oil industry's historical power and influence helps to place the struggle for clean water by the Standing Rock Sioux Tribe in context, for they were not simply fighting one oil company but instead an economic and cultural system "scaffolded" (as Jackson, 2010, puts it) by oil at its very core. In addition, the Tribe also found itself going up against what I term "deep media."

Deep media

When I first began researching scholarly conceptions of the "deep state," I continued to notice significant parallels between the characteristics attributed to the idea of a shadow government and those of the U.S. commercial media system. Like a shadow government, what I am terming *deep media* first recognizes that there is a highly visible media that audiences can engage with directly: they can read the articles, identify the journalists who wrote them,[12] and recognize broad patterns in coverage coming from individual media outlets and their owners. President Trump provides an excellent example of this when he calls out the owner of the *Washington Post* (Amazon CEO Jeff Bezos), the myth of the "failing" profits of the *New York Times*, or the politically "biased" media personalities hosting MSNBC's *Morning Joe*. Ownership, authorship, the superficial commercial logic, and potential "bias" of the news are all aspects of the media that most media audience members understand, regardless of whether or not they like, trust, or even regularly consume the mainstream media.

Behind all this, however, is a more obscured media system, one – like deep state – run with considerations of money, power, and influence at the forefront. *Deep media*, as I conceive of it here, has significant ties to government structures and practices; maintains close connections with not only the fossil fuel industry but corporate interests on a broad scale; demonstrates a strong gatekeeping function; challenges social equilibrium and democracy by being increasingly rooted in militarization and the *military-industrial complex*[13]; exhibits deep structural deficiencies due to its for-profit architecture and organization; and contains its own institutional inertia that is largely independent of individual actors (e.g., media owners, editors, and journalists) and their day-to-day actions.

Regarding the connections to government and the military, the numerous closed-door meetings held between the Bush administration and media owners and executives in the period leading up to the Iraq War attest to this idea of impactful, behind-the-scenes control, especially in the lack of critical journalism in the run-up to the war when the media industry worked closely with the government and other corporations to promote the war (Berman, 2006; Kumar, 2006; McChesney, 2008). The subtle – and often not so subtle – Hollywood films promoting the American military as part of Lenoir's (2000) "military-entertainment complex" – is another example. Embedded journalism – where major news outlets traveled and worked with the U.S. military, submitting their work to military and government agencies before they were published back at home – has been used as an illustration of the too-close relationship between the news media (Kumar, 2006). Other scholars have made these same observations, some of them going back almost half a century (Schiller, D., 2011; Mosco, 1986; Schiller, H. 1969). The often uncritical use of official government sources in professional news stories fits within this pattern too: powerful, well-connected sources are a journalist's "bread and butter," and thus antagonizing them is considered dangerous (Christians et al., 2011).

Is it a problem to accuse the news media of corporate bias? What is the push-back to this? As McChesney (2008, p. 35) observes, "establishing if there actually *is* a pro-corporate bias in the news is not an easy task and has been a source of more than a little controversy over the years. Although studies show the topic of corporate power is virtually unmentioned in U.S. political journalism, it is highly controversial to accuse journalism of a pro-corporate bias."

The media industry's ties with the fossil fuel industry are best revealed when it comes to environmental issues, and specifically with climate change. Holmes (2009) notes that "the campaign to cast doubt on climate change has been highly successful, and examples of the media's penetration by fossil fuel industry front-people abound" (p. 95). Media corporations often do not reveal their ties to other industries, including the fossil fuel giants, but the influence of the corporate profit motive on the media is persistent (Oreskes & Conway, 2011; Schudson, 2011; Rowell, 2007; Stauber & Rampton, 2002; Bagdikian, 1997). During the writing of this book, Meredith Corp, a U.S. media group, has purchased media company Time, which includes the *Time* magazine founded by publisher Henry Luce. A large portion of the money – $650 million – used to make the purchase came from fossil-fuel industry giant Koch Industries run by two brothers: Charles and David Koch. The crux of the problem, as former *Time* editor Charles Alexander (2017) phrased it in an article in *The Nation*, is not that the Koch brothers are politically conservative; rather it is that

> for decades, the Kochs [have] financed a campaign of disinformation designed to convince the public and politicians that climate change is nothing to worry about. In fact, any reputable climate scientist will tell you that global warming is the second-greatest danger to the human race, trailing only nuclear weapons.

Given the persistent ties between the media and fossil-fuel industries, perhaps it is not surprising that *Media Matters*, a not-for-profit media watchdog group, discovered that when Robert Bryce, a senior figure at the Manhattan Institute (which has received funding from ExxonMobil and the Koch Family Foundation) wrote pieces for the press in 2011, none of them disclosed his ties to industry. It was not simply conservative news outlets like Fox that failed to disclose Bryce's industry ties, but also left-leaning outlets like the *Huffington Post*, *Politico*, the *New York Times*, and CNN (Fong et al., 2011). As Oreskes and Conway make clear in their widely cited work *Merchants of Doubt*, the actions by large, powerful industries like the tobacco and fossil fuel industries means that the U.S. news media were slow to report critically on the global warming debate, despite evidence decades ago from the U.S. National Academy of Sciences that "there was no reason to doubt that global warming would occur from man's use of fossil fuels" (2011, pp. 242–243).[14]

Thus, another facet of *deep media* that is similar to that of the deep state (per Ambinder & Grady, 2013) is that it demonstrates a strong gatekeeping function through the withholding of key information from the public. Bagdikian (2004) refers to this as the "private ministry of information," where corporations and the government exert a heavy influence on what the media choose to report – and what they choose to omit from the public eye. While some scholars believe that secrets guarded by a deep state are necessary for national defense, it is less clear that a *deep media* system always extends those same benefits to the public. Indeed, if it is the case that "we can judge the quality of a democracy by the secrets it keeps" (Ambinder & Grady, 2013, p. 3), then the natural follow-up question would be to ask what information is withheld by the media when it comes to global issues like climate change, or environmental justice movements like #NODAPL, and what detriment exists in that concealment?

As noted, then, *deep media* refers more to the structural flaws inherent in the U.S. media system and less to conspiratorial activities based on the political bias of individual journalists (or corporate news owners).[15] That is, the concept has less to do with the biases of individual journalists or media owners, but instead has more to do with the structural flaws of the media landscape itself. Along these lines, Boykoff and Boykoff (2007) write that

> the rhythms and rituals of journalism do not simply cohere into a static structural factor; rather, they are built and buttressed by the everyday prac-titioners of journalism: reporters and editors who are enmeshed in political and professional discourses and normative orders.
>
> *(p. 1201)*

Similarly, Hall et al. (1978) observe that the architectural flaws in the news media represent "a set of structural imperatives" instead of "an open conspiracy with those in powerful positions" (p. 60). This is not to say that there is little merit in identifying and examining individual power players that shape financial and polit-ical trends in the U.S.: as one example, Jane Mayer's *Dark Money* (2016) provides an intriguing exploration of how billionaire Koch brothers and other powerful industry players have shifted the American political landscape more towards the "radical right." However, like a shadow government, *deep media* has its own logic independent of any one individual industry player, and thus is more focused on the structural vulnerabilities created by commercialism and corporate interests.

Michael Schudson (2011) deftly acknowledges as much, observing that there are certain structures and practices woven into the industry as a whole, making the apparent biases less driven by individuals and more systematic. I would agree, and contend that understanding these structural flaws lays the foundation to help explain the media coverage (and initial media blackout) by news media regarding the Dakota Access Pipeline struggle. Thus, I return to the idea of *deep media* in later chapters when I analyze media coverage of the Dakota Access Pipeline and the history of the law and order approach in the U.S.

At this juncture, it should become clear to anyone who studies the media that with *deep media* I am technically expanding and providing a new vantage point to view a phenomenon already recognized in political economy scholarship – namely, that the mainstream media are owned and run by the wealthy and powerful, those who work in tandem with government, military, and with other corporations (including those in the oil industry) behind the scenes to gain control, exert influence, and win power. Hall et al. (1978) recognized decades ago that the media have a "structured relationship to power" (p. 59). More recently, McChesney (2008) has alluded briefly to an "underground press" that not only exhibited the weaknesses I have described above as deep media but that existed during a time often considered journalism's "golden age." Kumar (2006) and Kellner (2004) have provided particularly good examples of what *deep media* might look like in their analysis of U.S. press involvement in the Iraq War. Kumar (2006) identifies what he calls a "mechanism of information control" represented by the U.S. government as well as a "for-profit giant conglomerate media system that lends itself to propaganda due to its structural limitations" (p. 49). Kellner (2004) observes that the "corporate press" has become the "arms of conservative and corporate interests which advance state and corporate agendas" (p. 31). Aside from furthering government and industry interests, the additional danger of this, Kellner contends, is that the news media then lose their public service component and instead participate in "undermining democracy" (p. 31).

In sum, the similarities between deep state and deep media are intriguing. Both of them demonstrate corporate ties, especially to the fossil fuel industry, creating a significant conflict of interest. Together they demonstrate a clear gatekeeping function regarding key information when it benefits government or corporate interest. As such, deep media as defined here overlaps, complements, and works in concert with the deep state on a deep foundational level, creating the potential endangerment of social equilibrium and democracy when it comes to social and environmental justice movements like #NODAPL. In the next section, I discuss more in depth the deep structural weaknesses of the news media system. In particular, I focus briefly on professional journalism, public relations, the commercial structure of the news, and the "Trump Bump" as part of my critique. Although there are several excellent political economy critiques, this discussion provides a useful update and synthesis of this scholarship on the path towards understanding which factors influenced media coverage of #NODAPL.

Power, money, and the media: the structural constraints of commercial journalism

As the preceding section makes clear, while the oil industry was a powerful player in the fight over the Dakota Access Pipeline, the media industry also can be considered influential. Wasko et al. (2011) identify the importance of applying a critical lens to the media

> precisely because the communications industries play a central double role in modern societies, as industries in their own right and as the major site of the representations and arenas of debate through which the overall system is imagined and argued over.
>
> *(p. 2)*

Following from this recognition, this section identifies contemporary critiques of the news media system so as to lay the foundation to understand press coverage of Standing Rock, from the initial media blackout to later coverage at the height of the struggle. I should note that this is not an intent to wade into ongoing debates in political economy as to whether or not the media industry is indeed concentrated (and if it is, why that matters), for others have made compelling arguments on this issue (see Bettig & Hall, 2012; Downing, 2011; McChesney, 2008; Andersen & Gray, 2008; Kunz, 2007; Bagdikian, 2004; Thomas & Nain, 2004; Herman & Chomsky, 2002; Miller, 2002). Here, I provide an evaluation of the U.S. journalism landscape through the lens of the economic tensions and structural flaws inherent in professional journalism. As Wasko (2014, p. 261) has observed, the "marketization," "commercialization," and "privatization" of the media landscape has been ongoing for several decades now – so much so that this period can be recognized as one of "cultural capitalism." What are the new norms of journalism in the period in which we find ourselves now? What are some of the structural weaknesses in the press that arise as a result of this? These questions provide the focal point for the next few sections.

Professional journalism: the idea of an "objective" news system

One of the primary focal points for political economy critiques of the U.S. press is professionalism journalism, a trend that grew out of the Progressive Era as a response to claims of corruption in the news media (Schudson, 2011; McChesney, 2008). Professionalism in journalism places primary emphasis on remaining "detached" and, in theory, unbiased. McChesney (2008) observes that

> the notion that journalism should be politically neutral, nonpartisan, professional, even 'objective,' did not emerge until the twentieth century. During the first two or three generations of the Republic such notions for

the press would have been nonsensical, even unthinkable. The point of journalism was to persuade as well as inform...The free press clause in the First Amendment to the Constitution was seen as a means to protect dissident political viewpoints.

(p. 26)

Kuypers (2013, p. 62) observes that the goal of objectivity effectively quelled the investigative journalism and "muckraking" to the point that objectivity became the "industry standard." Although "objectivity" is the hallmark of professional journalism, it is important to distinguish between 'impartiality' (rooted in pluralism and relativism with no sense of an immutable truth) and 'objectivity' (which holds that the world can be and should be described accurately) (Holmes, 2009, p. 99). That is to say, impartiality assumes no concrete basis for truth, which then reduces any claim to simply another competing truth in the marketplace of ideas; objectivity, on the other hand, makes a claim to tell an absolute truth – which then invites the question of which journalist, reporter, or media outlet can ever make a claim to be truly value free. The idea of a value-neutral press creates a serious challenge, for even the attempt to appear non-partisan can carry its own bias: as McChesney (2008) notes, "journalism cannot actually be neutral or objective, and unless one acknowledges that, it is impossible to detect the values at play that determine what becomes news, and what does not" (p. 30). This is a particularly important point when considering the media blackout in the initial stages of the struggle over the Dakota Access Pipeline (discussed more in Chapters Three and Four), for it is clear that the professional news industry, especially in the U.S., did not feel the Standing Rock Sioux Tribe's fight merited coverage at all.

Schudson (2011) seems to agree with the standing critiques of professional journalism while identifying additional issues, noting that journalistic professionalism is largely episodic (focused on events), focuses on strategy and tactics rather than context,[16] leans towards reporting on negative news, tries to appear objective at all costs, and relies heavily on official sources.[17] When it comes to office sources in particular, McChesney (2008) pinpoints the problem: the creation of unequal power relationships, where those in power get to prioritize certain aspects of a story, or even determine the story itself, which is similar to Blyskal and Blyskal's (1985) notion of the distorting "fish-eye lens." Speaking to corporate control of not just the media but American culture, Hedges (2009) forcefully argues that:

> If you are a true journalist, you should start to worry if you make $5 million a year. No journalist has a comfortable, cozy relationship to the powerful. No journalist believes that serving the powerful is a primary part of his or her calling.
>
> *(p. 169)*[18]

As a result of the pressing issues facing journalism, Schudson (2011, p. 49) sums it up thusly: "the problem with the press is professionalism, not its absence."

Public relations, advertising, and the profit motive in professional journalism

Related to the idea of corporate or government influence through official sources is the infiltration by public relations (PR) firms into the news media:

> Another weakness built into professional journalism as it developed in the United States was that it opened the door to an enormous public relations industry that was eager to provide reporters with material on their clients. Press releases and packets came packaged to meet the requirements of professional journalism, often produced by former journalists. The point of public relations (PR) is to get the client's message in the news so that it looks legitimate: The best PR is never recognized for what it is. Although reporters generally understood the dubious nature of PR, and never embraced it, they had to work with it to get their job done.
>
> *(McChesney, 2014, p. 19)*

The structural vulnerabilities created by professional journalism have made it more susceptible to commercialism and profit motive. This is less a blanket statement about the quality of American journalism and more of a recognition of the tensions presented by corporate ownership of the media as well as the media's reliance on advertising revenue for its survival. Downing (2011) provides an excellent treatment of what is often colloquially referred to as the "Hutchins Report."[19] In the document, Hutchins laid the groundwork for a socially responsible press while expressing his fears regarding the conflict of interest created by financial interest in the public-service function of the news media, writing that "The major part of the nation's press is large-scale enterprise, closely interlocked with the system of finance and industry; it will not without effort escape the natural bias of what it is" (1947, 129–130, in Downing, 2011, p. 145). Downing notes that Hutchins was wary in particular of the "oligopolistic tendency of the US media" and the enmeshment of the press with industry and economic concerns (p. 145). Hutchins' wariness was centered on the debate around democracy:

> These formulations adumbrated a continuing theme in US debate concerning media ownership concentration: the threat to democracy from a purely commercially driven media system, and the threat to effective market competition. In this version, the former is given more weight than the latter.
>
> *(Downing, 2011, p. 145)*

Schauster et al. (2016) echo the recognition that journalism should maintain a wall between itself and corporate influence: "The established goal of social responsibility theory is to ensure that business and government cannot influence journalism, effectively meaning that the press can act as autonomous watchdog of power" (p. 1420). What Schudson (2011) terms the "drive for profit" harms journalism in myriad ways, including reducing personnel and cutting the commitment to investigative journalism (which is expensive and time-consuming). The economic logic of professional, commercial journalism dictates that while investigative, critical stories decrease, sensationalistic "fluff" pieces rise in number:

> Since the early 1980s, commercial pressure has eroded much of the autonomy that professional journalism afforded newsrooms, and that provided the basis for the best work done over the past 50 years. It has led to a softening of standards such that stories about sex scandals and celebrities have become more legitimate, because they make commercial sense: They are inexpensive to cover, attract audiences and give the illusion of controversy without threatening anyone in power.
>
> *(McChesney, 2014, p. 19)*

Another way to frame the discussion around profit motive in U.S. journalism is through the *resource dependence perspective*,[20] which

> proceeds from the indisputable proposition that organizations are not able to internally generate either all the resources or functions required to maintain themselves, and therefore organizations must enter into transactions and relations with elements in the environment that can supply the required resources and services.
>
> *(Aldrich & Pfeffer, 1976, p. 83)*

Put another way: "We can better understand how media producers… behave by understanding how their resources are allocated" (Grossberg et al., 2005, p. 74). The *resource dependence perspective* thus acknowledges the external economic influences exerted on journalism, which, in this context, is less individual subscriber dollars and more advertising revenue. A well-known (in political economy circles) example of the resource dependence model comes from journalist and editor Bill Kovach after he was hired by the *Atlanta Journal Constitution* newspaper in the late 1980s. The paper hired Kovach to give credibility to the paper through investigative journalism and critical reporting, but after he investigated a series of local banks and one of the largest employers in Atlanta, "management of the Atlanta newspapers had been under pressure to rein in Mr. Kovach because of his aggressive coverage of the business community" (Scardino, 1988). Thus, although the newspaper yearned for more journalistic validity, it refused to risk alienating its funding sources and pressured Kovach to resign.

Public Broadcasting in a Commercial Media Landscape

Lest one think that public broadcasting – or "non-profit broadcasting" – is immune from commercial influence, it is important to note that the federal government only provides a small percentage of the revenues for these stations, money that they must make up with corporate sponsorship or viewer contributions. This creates a kind of paradox: if public broadcasters refuse corporate money, they have a difficult time staying afloat; but if they do take the money, they have a difficult time maintaining their public service commitment (McChesney, 2008). Perhaps it is for this reason that the "Charlie Rose" show on *Public Broadcasting Service* (PBS) was underwritten by the NewsCorp conglomerate and sponsored by Coca Cola (Kaplan, 2009). It is difficult to come to any other conclusion except that the non-profit media sector has been under commercial and political attack for many years, with the most recent example being the attempt, early in 2017, by President Trump to remove all federal funding for public broadcasting.

Schudson (2011) notes that "the boundaries delineating for-profit, public, and nonprofit media have blurred, and cooperation across these models of financing has developed" (p. 212). In a particularly candid early interview with the *New York Times* (Barboza, 1995), one NPR studio manager put it succinctly:

> "We go ahead and push the envelope a little," says Doug Myrland, general manager at KPBS, the NPR member station in San Diego, which has used longer credits and corporate theme music to more than double its underwriting earnings in the last year. "Stations have gotten serious about corporate sponsorship. *Value-added is now part of our vocabulary.*" [emphasis added]

While caution could be advised in claiming that public broadcasting is inherently corporate or commercial in the same way as Disney-owned ABC, it is important to at least recognize some of the pressures faced by non-profit news media in the U.S.

Trump and the (temporary?) financial revival of watchdog journalism

McChesney (2014) makes a forceful argument: "In my view, the evidence suggests that US-style professional journalism under capitalist management offers no hope for the people of the United States, let alone anywhere else on the planet" (p. 12). The quality of journalism suffers even more due to the loss of advertising and subscriber dollars. The trend, of course, is a sharp shift away from hard-copy newspapers sold to individual paying subscribers and towards relatively open access of information on the Internet. Couldry and Turow (2014) observe that this shift is even more significant than it seems, for advertisements don't really follow companies anymore, but instead they follow the media consumer. Because of this, McChesney recognizes an important trend in the news industry: as subscriber and corporate dollars have abandoned it, the news sector has shrunk significantly. When he traveled around the country visiting various journalists, he learned that morale (and employment numbers) of those in the industry is incredibly low. He also doesn't see that much will change in this regard in journalism as a whole.[21] When Trump began to run for president, however, the fortunes of many news media outlets began to change, and thus it is worth assessing how the landscape of journalism has changed in response to this.

A democratic theory of the media holds that the press should: 1) be a watch-dog for the public and hold those in power (or those who wish to be in power) accountable; 2) be able to discern truth from lies; 3) "regard the information needs of all people as legitimate; if there is a bias in the amount and tenor of coverage, it should be towards those with the least amount of economic and political power"; and 4) present a balanced range of perspectives and opinions on important issues future and present (McChesney, 2014, p. 13).[22] While commer-cialization of the media has significantly hampered the ability of the press to ful-fill this normative ideal for many decades, the presidential candidacy – and then election – of Donald Trump has changed the landscape somewhat. First, one can consider cable news: as Derek Thompson (2016) with *The Atlantic* observes:

> Since Trump descended that escalator last June, cable news' fortunes— particularly CNN's—have been ascendant. Total primetime viewership for the three channels grew by 8 percent in 2015, and profits soared by about a fifth at both CNN and Fox News. Trump may be destroying U.S. democratic norms, but he appears, for the moment, to be one big beautiful orange life raft for the flagging cable news business.

This phenomenon is now colloquially referred to as the "Trump Bump" (or the "Trump Effect"), where subscriptions to news media and number of digital "eyeballs" have grown since Trump's rise to power. It is clearly not just cable news that has benefited: subscriptions to national newspapers and magazines also have risen. The *New York Times*, which Trump likes to argue is finan-cially "failing," actually reported higher quarterly earnings than expected in 2016 and 2017. The *Times* CEO Mark Thompson, in an interview on CNBC, reported that "We added an astonishing 308,000 net digital news subscrip-tions, making Q1 the single best quarter for subscriber growth in our history" (Wojcik, 2017). Thompson, subtly alluding to the Trump Effect, noted that the *Times* had "some rocket fuel in the subscription business in recent quar-ters." Other national publications have fared similarly. Trump's sharp criticism of *Vanity Fair* magazine in late 2016 led to the largest number of new subscrip-tions for the magazine's parent company Conde Nast in a single day (Politi, 2016).[23] Worth noting is that the Trump Bump does not just include journal-ism but impacts some progressive causes as well, including the *American Civil Liberties Union* (whose donations page crashed with all the postelection activity) and the family planning clinic *Planned Parenthood*.

Trump's antagonism towards the media, coupled with his own rising unpopu-larity, appears to have created an interesting contradiction that could be called the "Trump Paradox": while he relentlessly attacks most news media as "fake," he also has bolstered their profits considerably due to his undeniable media draw. As Feldman (2016) notes, "hate watching" Trump has become a profitable enterprise. Thus, newspapers and Trump currently have a highly contentious yet almost symbiotic relationship, one where Trump routinely attacks the authenticity of mainstream news, yet they still cover him, and he continues to ensure their strong profit margin.

News outlets appear to be well aware of the Trump Effect: *Vanity Fair* added a tagline to its digital subscriptions page that read "The Magazine Trump Doesn't Want You to Read." The *Washington Post* added a line at the bottom of its articles that read "Democracy Dies in Darkness." Reflecting this awareness, Callum Borchers (2017) at the *Washington Post* went so far as to pose this question in the title of his article: "Is Donald Trump Saving the News Media?"

Although Trump's impact on national news media can be seen as both positive (in terms of a shift to stronger watchdog journalism) and substantial (in terms of both subscription dollars and eyeballs), local papers (the ones that don't have a strong national or DC section), are either hovering at the same level of subscriptions or are in a steep freefall:

> local publications…are still suffering through the industry's long decline and need to retain subscribers who are sympathetic to Trump. Consider McClatchy Co., which owns about 30 papers, including the Miami Herald. Its shares have plummeted 31 percent since Election Day. Subscriptions have barely budged.
>
> *(Smith, 2017)*

Thus, the financial revival of newspapers seems limited to the national press and may also be limited in terms of time frame: Trump will not be president forever, and it is worth asking if audiences will still want to subscribe to journalism (and be politically engaged) when their favorite *bête noire* has faded from view.

Consider this as well: the news media have experienced a surge in subscriptions because they have reported on Trump's foibles as a *political* representative. The press's eagerness to report on government malfeasance is in line with standard professional journalism practice. But what about corporate misconduct? When journalist Charles Lewis created the *Center for Public Integrity* in the 1990s, his intent was to (along with his assembled team of journalists) create a series of investigative journalism reports that he would then send on to news outlets for wider dissemination. What he found was that the news media were keen to cover his stories about government misbehavior but not corporate corruption (McChesney, 2008). So, while it is clear that watchdog journalism may be on the rise in relation to Trump, it may be circumscribed by the type of news outlet (national), by the length of time Trump is in office, and also by the type of story (government or corporate). That Trump straddles both arenas makes this even more interesting: the media have been quick to report on some of his questionable business practices, but will this critical media eye extend to other corporate activity as well, especially those involving the environment and human rights?

This is important when considering Indigenous struggles over the Dakota Access Pipeline.

As this brief section on the problems and trends has made evident, journalism in the U.S. finds itself in flux and it is anyone's guess where it will go once Trump is no longer simultaneously antagonizing and energizing the national news media in his capacity as U.S. president. However, it is safe to say that the commercial logic and corporate ownership of the media is likely to continue for the foreseeable future. In addition to political shifts, it is important to consider how technological changes have significantly impacted journalism, especially in the sense that: 1) television and newspapers have ceased to be the dominant source of information in the U.S.; 2) traditional media sources have lost much of their information gatekeeping function; 3) funding sources have significantly shifted; and 4) "citizen journalism" is on the rise. Covered more in Chapter Eight in relation to environment reporting and #NODAPL, the influence of the Internet on journalism (especially on environmental issues) is difficult to overestimate: as Sachsman et al. (2010) observe, "the future of environment reporting has more to do with the future of news than it does with the future of newspapers" (p. 179).

The cost of having "fossil fuel values": #NODAPL, human rights, and ecology

In the preceding sections I have noted how powerful the players are who are involved in the struggle over the Dakota Access Pipeline, including the oil industry and its corporate backers as well as the media system. However, in the case of #NODAPL, there was an emerging power that could not be ignored: the Standing Rock Sioux Tribe. As the outpouring of celebrity and popular support made clear, the Tribe appears to have won (as one newspaper phrased it) the "public relations" battle. Most significant, they seemed to have won this particular campaign *despite* the powerful, well-established monied interests behind the pipeline – Big Oil, the state government, and many mainstream media outlets. Klein (2014) refers to the uphill battle Indigenous groups face as "might vs rights," because while the Tribe effectively made the case for human rights, Energy Transfer Partners retains its money and influence. Along these lines, Klein goes on to note that

> in Canada, the United States, and Australia, these rights are not only ignored, but Indigenous people know that if they try to physically stop extractive projects…they will in all likelihood find themselves on the wrong side of a can of pepper spray – or the barrel of a gun. And while the lawyers argue the intricacies of land title in court, buzzing chainsaws proceed to topple trees that are four times as old as our countries, and toxic fracking fluids seep into the groundwater.
>
> *(pp. 377–378)*

The reason she gives for why industry has often gotten away with this is what she calls "raw political power" (p. 378), as well as help from the "corporate-owned media." To restate a point made above, however, the Tribe was not simply going up against a few corporate interests: to simply identify the significant power of the oil industry and the institutionalized flaws of the commercial media giants is to ignore the structural realities of contemporary life on this planet, a life that is completely dependent on fossil fuels. Morris (2015) identifies several different types of value systems, including those of "foragers," "farmers," and "fossil fuel users."[24] What he means by this is what any method of "energy capture" (which-ever one is used most widely and commonly) is so important that it determines a society's values. This idea of "fossil fuel values" is echoed in Wesley Jackson's (2010) *Consulting the Genius of the Place*, where he observes that our contem-porary civilization is inextricably predicated on the "five exhaustible and rela-tively nonrenewable carbon pools found in soil, trees, coal, oil, and natural gas" (p. 14). Because of this, Takach (2016) writes that, while the end goal of "end-less growth" is unsustainable from a moral, environmental, and scientific stand-point, the mindset is still popular: "inertia is a formidable force in any system, and the privileged, relative few who are well served by business as usual have invested tanker-loads in riding the one-trick pony of fossil-fuel extraction into the ground, consequences be damned" (p. 21).

The concept of *deep media*, which recognizes the structural flaws created by the economic undergirding and profit motive of the media industry, is useful to understand the press coverage of Standing Rock. The reason for this is that the best explanation for why the press initially ignored the struggles of the Standing Rock Sioux Tribe comes not from a *micro* perspective (e.g., individual news outlets and their owners) but instead a broader vantage point that explores why the media industry as a whole chose to overlook the issue. When one begins to examine that, the connections between the media industry and other corporate and state influences begin to emerge more clearly.

Although the mainstream media initially created a blackout on #NODAPL, the Standing Rock Sioux Tribe distinguished themselves because they changed the contours of the game a bit. That is, they managed to bend the existing rules set in place by corporate interests in order to shift public opinion in their favor through an effective use of social media and working with more traditional forms of journalism (like the *New York Times* in the U.S. and *The Guardian* in the UK). Part of their success is that they created meaningful rhetoric like "Mni Wiconi" ("Water is Life") that resonated with the broader public. Solnit (2016) notes that

> Any involved in the climate movement see it as a human rights movement
> or a movement inseparable from human rights. Indigenous people have
> played a huge role, as the people in many of the places where extracting
> and transporting fossil fuel take place, as protectors of particular places and
> ecosystems from rivers to forests, from the Amazon to the Arctic, as people

with a strong sense of the past and the future, of the deep time in which short-term profit turns into long-term damage, and of the rights of the collective over individual profit.

The Standing Rock Sioux Tribe effectively tapped into a growing environmental mindset respecting the Indigenous perspective and did so at a time of shifting beliefs about fossil fuels and environmental issues as well as a changing media landscape. This book steps into this struggle to examine the role that media played in covering environmental justice issues like #NODAPL around the world at a time of heightened concern about climate change and human rights. The next chapter begins to narrow its assessment of the media landscape to explore how a commercially driven press system covers environmental issues, including environmental justice.

Notes

1 Sampson (1975) identifies the seven: Exxon/Esso, Shell, BP, Gulf, Texaco, Mobil and Social (later Chevron).
2 Yergin (2008) actually identifies an eighth "sister."
3 Juhasz cites Obama's speech after winning the Iowa Caucus in 2008, where he pledged to end "the tyranny of oil."
4 Intriguingly, Mitchell (2011) sees the link between oil and democracy as resting on the oil industry's infrastructure – the production of (or the "apparatus of") the oil industry. Mitchell refers to this link not as democracy and oil but "democracy *as* oil" (emphasis in original) – a provocative connection.
5 Saudi Arabia provides a good example of this recognition: due to its concern about its own oil consumption, it is engaged in numerous solar projects around the country as a way to provide power for its citizens (Ball, 2015). Forward-thinking executives at British Petroleum attempted to rebrand the corporation as "Beyond Petroleum" and invested in renewable energy until a significant divestiture from it in 2011 (when the company decided fossil fuels were more profitable in the near future).
6 Jane Mayer (2016), in her impressive tome *Dark Money*, makes clear that the jury is still out on whether or not the oil industry represents a true oligarchy: "Some might dispute that American oil, gas, and coal magnates meet the dictionary definition of a small, privileged group that effectively rules over the majority" (p. 200). But, she notes, it is "indisputable" that these powerful players funded and organized effective counter-arguments against climate change, thereby slowing the adoption of climate change mitigation strategies.
7 Having stated this, the foundation of the argument made by Princen et al. (2015) is a recognition of what they term a "perceptual shift" away from fossil fuels, a shift that they claim can be seen gaining momentum worldwide.
8 According to Lofgren (2016) and Scott (2015), the phrase "deep state" was coined in Turkey in the late 1990s as a way to refer to U.S. covert operations in the country.
9 Intriguingly, much has been made recently of this "shadow government," especially in contemplation of what the established government insiders and latent power structure will do to the political rogue Trump. A simple Internet search reveals numerous conversations (most emanating from conservative news outlets) about how the Washington political elite will stop Trump from achieving his goals. Trump may well represent a part of that influence as he now straddles both the political and corporate arenas: as many news outlets reported, he owned as much as a million dollars of stock in Energy Transfer Partners in 2014, which he reduced to under $50k in 2015.

Many have voiced concerns about how his investments and corporate status impact decisions he has made on the Dakota Access Pipeline project.

10 Berman (2006) supplies an excellent, in-depth discussion of how the U.S. was involved in removing democratically elected Iranian Premier Mossadegh from power to prevent him from taking control of his country's oil supply back from Great Britain.

11 This symbiotic relationship is reinforced by the fact that, according to Chastko (2004), Alberta's oil sands have been predicted to create about 50 percent of the total "crude oil output" for Canada.

12 In the case of PR pieces posing as journalism, identifying the original author can be a challenge.

13 Peter Scott's work *The American Deep State* (2015) discusses the military-industrial complex as part of the deep state in depth.

14 In addition, President George H.W. Bush already had made a commitment to the U.S. Framework Convention on Climate Change in 1992 (Oreskes & Conway, 2011).

15 Schudson (2011) makes a similar observation in his distinction between bias (individual and unpredictable) and framing (institutional and thus more predictable).

16 McChesney (2008) argues that journalism "tends to avoid contextualization like the plague" because to do so would "commit the journalist to a definite position and enmesh the journalist…in the controversy professionalism is determined to avoid" (p. 33).

17 McChesney (2008), like Schudson, believes that professional journalism relies on official sources so heavily as a way to "remove the controversy connected with the selection of stories" and continue to appear objective and fair (p. 31).

18 In this case, Hedges recounts an interview with Jim Cramer (host of CNBC's Mad Money) by Jon Stewart on the *Daily Show*, where Stewart accuses Cramer of being complicit in the U.S. Great Recession in 2008 by not being critical of the questionable practices of large financial institutions.

19 The full name is *A Free and Responsible Press: A General Report on Mass Communication: Newspapers, Radio, Motion Pictures, Magazines, and Books.*

20 Turow (1984) is most commonly cited for his adaptation of the *resource dependence perspective* but his treatment of it is clearly based on this earlier text by Aldrich and Pfeffer (1976).

21 Although he holds this perspective, it's important to note that morale in environmental reporting is higher per Sachsman et al. (2010).

22 McChesney cites Christians et al. (2008) as the ultimate origin of the democratic theory of media.

23 Although Politi notes that Trump did not mention it by name, likely he was angered by a December 2016 article by Tina Ngyuen that wittily excoriated one of Trump's restaurants: "Trump Grill Could Be the Worst Restaurant in America."

24 Morris, however, does believe that we are ultimately headed into a "post-fossil fuel stage," although he recognizes the inertia created by the fossil fuel system.

References

Aldrich, H. E., & Pfeffer, J. (1976). Environments of organizations. *Annual Review of Sociology, 2,* 79–105.

Alexander, C. (2017, November 22). Don't let the Koch brothers buy 'Time' magazine. *The Nation.* Retrieved from https://www.thenation.com/article/dont-let-the-koch-brothers-buy-time-magazine/

Ambinder, M., & Grady, D. B. (2013). *Deep state: Inside the government secrecy industry.* Hoboken, NJ: Wiley.

Andersen, R., & Gray, J. (2008). *Battleground: The media*. Westport, CT: Greenwood Press.

Bagdikian, B. H. (2004). *The new media monopoly*. Boston, MA: Beacon Press.

Ball, J. (2015, Jul 1). Why the Saudis are going solar. *The Atlantic Monthly, 316*, 72.

Barboza, D. (1995, December 7). The 'enhanced underwriting' of public broadcasting is taking a more commercial flair. *New York Times*. Retrieved from http://www.nytimes.com/1995/12/27/business/media-business-advertising-enhanced-underwriting-public-broadcasting-taking-more.html?mcubz=3

Berman, M. (2006). *Dark ages America: The final phase of empire*. New York: W.W. Norton & Co.

Bettig, R., & Hall, J. (2012). *Big media, big money: Cultural texts and political economics*. New York: Rowman & Littlefield.

Blyskal, M. H., & Blyskal, J. (1985). *PR: How the public relations industry writes the news*. New York: W. Morrow.

Borchers, C. (2017, February 6). Is Donald Trump saving the news media? *Washington Post*. Retrieved from https://www.washingtonpost.com/news/the-fix/wp/2017/02/06/is-donald-trump-saving-the-news-media/?utm_term=.86851b4504ef

Boykoff, M. T., & Boykoff, J. M. (2007). Climate change and journalistic norms: A case-study of US mass-media coverage. *Geoforum, 38*(6), 1190–1204. 10.1016/j.geoforum.2007.01.008

Christians, C. G., Fackler, M., Richardson, K., Kreshel, P., & Woods, R. (2011). *Media ethics: Cases and moral reasoning* (9th ed.). Boston, MA: Allyn & Bacon.

Couldry, N., & Turow, J. (2014). Advertising, big data, and the clearance of the public realm: Marketers' new approaches to the content subsidy. *International Journal of Communication, 8*, 1710–1726.

Downing, J. (2011). Media ownership, concentration, and control: The evolution of debate. In J. Wasko, G. Murdock, & H. Sousa (Eds.), *The handbook of political economy of communications* (pp. 140–168). Malden, MA: Wiley-Blackwell.

Engelhardt, T. (2014). *Shadow government surveillance, secret wars, and a global security state in a single-superpower world*. Chicago, IL: Haymarket Books.

Feldman, J. (2016, August 25). The science of 'hate-watching': What Trump is teaching TV execs. *AdAge*. Retrieved from http://adage.com/article/guest-columnists/hate-watching-trump-teaching-tv-execs/305601/

Fong, J., Theel, S., & Fitzsimmons, J. (2011). *Who is Robert Bryce? Media Matters*. Retrieved from https://www.mediamatters.org/research/2011/10/07/who-is-robert-bryce/181888

Gareau, F. (2002). *The United Nations and other international institutions: A critical analysis*. Chicago, IL: Burnham, Inc.

Goodman, A. (2017, May 22). PBS's Tavis Smiley interviews Amy Goodman. *Tavis Smiley Show*. Retrieved from https://www.democracynow.org/2017/5/22/watch_tavis_smiley_interviews_amy_goodman

Grossberg, L., Wartella, E., Whitney, C., & Wise, J. (2005). *Mediamaking: Mass media in a popular culture* (2nd ed.). Thousand Oaks, CA: Sage Publications.

Hall, S., Critcher, C., Jefferson, T., Clarke, J., & Roberts, B. (1978). *Policing the crisis: Mugging, the state, and law and order*. New York: Holmes & Meier.

Hedges, C. (2009). *Empire of illusion: The end of literacy and the triumph of spectacle*. New York: Nation Books.

Herman, E. S., & Chomsky, N. (2002). *Manufacturing consent: The political economy of the mass media*. New York: Pantheon Books.

Holmes, T. (2009). Balancing acts: PR, "impartiality," and power in mass media coverage of climate change. In T. Boyce & J. Lewis (Eds.), *Climate change and the media* (pp. 92–101). New York: Peter Lang.

Jackson, W. (2010). *Consulting the genius of the place: An ecological approach to a new agriculture.* Berkeley, CA: Counterpoint Press.

Juhasz, A. (2008). *The tyranny of oil: The world's most powerful industry—and what we must do to stop it.* New York: William Morrow.

Kaplan, D. (2009, Sept 28). Why business loves Charlie Rose. *Fortune Magazine.* Retrieved from http://archive.fortune.com/2009/09/25/magazines/fortune/charlie_rose.fortune/index.htm

Kellner, D. (2004). The media and the crisis of democracy in the age of Bush. *Communication and Critical/Cultural Studies, 1*(1), 29–58. 10.1080/1479142042000180917

Klein, N. (2014). *This changes everything: Capitalism vs. the climate.* New York: Simon & Schuster.

Kumar, D. (2006). Media, war, and propaganda: Strategies of information management during the 2003 Iraq war. *Communication and Critical/Cultural Studies, 3*(1), 48–69. 10.1080/14791420500505650

Kunz, W. M. (2007). *Culture conglomerates: Consolidation in the motion picture and television industries.* Lanham, MD: Rowman & Littlefield.

Kuypers, J. A. (2013). *Partisan journalism: A history of media bias in the United States.* Lanham, MD: Rowman & Littlefield.

Lofgren, M. (2016). *The deep state: The fall of the constitution and the rise of a shadow government.* New York: Viking.

Mayer, J. (2016). *Dark money: The hidden history of the billionaires behind the rise of the radical right.* New York: Doubleday.

McChesney, R. W. (2008). *The political economy of media: Enduring issues, emerging dilemmas.* New York: Monthly Review Press.

McChesney, R. (2014). The struggle for democratic media: Lessons from the north and from the left. In C. Martens, E. Vivares, & R. McChesney (Eds.), *The international political economy of communication: Media and power in South America* (pp. 11–30). London: Palgrave Macmillan.

Miller, M. C. (2002). What's wrong with this picture? *Nation, 274*(1), 18–22.

Mitchell, T. (2011). *Carbon democracy: Political power in the age of oil.* New York: Verso.

Morris, I. (2015). *Foragers, farmers, and fossil fuels: How human values evolve.* Princeton, NJ: Princeton University Press

Mosco, V. (1986). New technology and space warfare. In J. Becker, G. Hedebro, & L. Paldan (Eds.), *Communication and domination: Essays to honor Herbert I. Schiller* (pp. 76–83). Norwood, NJ: Ablex.

Nicholson, B. (2017, May 30). Dakota access pipeline, law officers had close relationship. *AP Worldstream.* Retrieved from https://search.proquest.com/docview/1903683443

Oreskes, N. & Conway, E. (2011). *Merchants of doubt: How a handful of scientists obscured the truth on issues from tobacco smoke to global warming.* New York: Bloomsbury Press.

Politi, D. (2016, December 17). Vanity Fair subscriptions soar after Trump blasts magazine on Twitter. *Slate.* Retrieved from http://www.slate.com/blogs/the_slatest/2016/12/17/vanity_fair_subscriptions_soar_after_trump_blasts_magazine_on_twitter.html

Princen, T., Manno, J., & Martin, P. (2015). *Ending the fossil fuel era.* Cambridge, MA: The MIT Press.

Rowell, A. (2007). Exxon's footsoldiers: The case of the international policy network. In W. Dinan & D. Miller (Eds.), *Thinker, faker, spinner, spy: Corporate PR and the assault on democracy* (pp. 94–116). London: Pluto Press.

Sachsman, D. B., Simon, J., & Valenti, J. M. (2010). *Environment reporters in the 21st century.* New Brunswick, NJ: Transaction Publ.

Sampson, A. (1975). *The seven sisters: The great oil companies and the world they shaped*. New York: Viking Press.

Scardino, A. (1988, November 5). Atlanta editor resigns after dispute. *New York Times*. Retrieved from http://www.nytimes.com/1988/11/05/us/atlanta-editor-resigns-after-dispute.html?mcubz=3

Schauster, E. E., Ferrucci, P., Neill, M. S., Wojdynski, B. W., & Golan, G. J. (2016). Native advertising is the new journalism. *American Behavioral Scientist, 60*(12), 1408–1424. 10.1177/0002764216660135.

Schiller, D. (2011). The militarization of US communications. *Communication, Culture and Critique 1*(1), 126–138. 10.1002/9781444395402.

Schiller, H. I. (1969). *Mass communications and American empire*. New York: A.M. Kelley.

Schudson, M. (2011). *The sociology of news* (2nd ed.). New York: W.W. Norton & Company.

Scott, P. D. (2015). *The American deep state: Wall street, Big Oil, and the attack on U.S. democracy*. Lanham, MD: Rowman & Littlefield.

Shrivastava, M. & Stefanick, L., (2015). Introduction: Framing the debate on democracy and governance in an oil-exporting economy. In M. Shrivastava & L. Stefanick (Eds.), *Alberta oil and the decline of democracy in Canada*. Edmonton, AB: AU Press.

Smith, G. (2017, July 10). Trump bump for president's media archenemies eludes local papers. *Bloomberg*. Retrieved from https://www.bloomberg.com/news/articles/2017-07-10/trump-bump-for-president-s-media-archenemies-eludes-local-papers

Solnit, R. (2016, September 12). Standing Rock protests: This is only the beginning. *The Guardian*. Retrieved from https://www.theguardian.com/us-news/2016/sep/12/north-dakota-standing-rock-protests-civil-rights

Stauber, J., & Rampton, S. (2002). *Trust us, we're experts!: How industry manipulates science and gambles with your future*. New York: Jeremy P. Tarcher/Putnam.

Tabuchi, H. (2016, November 7). Environmentalists target bankers behind pipeline. *New York Times*. Retrieved from https://www.nytimes.com/2016/11/08/business/energy-environment/environmentalists-blast-bankers-behind-dakota-pipeline.html?mcubz=3

Takach, George F (George Francis). (2016). *Scripting the environment: Oil, democracy and the sands of time and space*. Cham, Switzerland: Palgrave Macmillan.

Thomas, P., & Nain, Z. (2004). *Who owns the media? Global trends and local resistances*. London: Palgrave.

Thompson, D. (2016, June 17). The 'Trump effect' on cable news. *The Atlantic*. Retrieved from https://www.theatlantic.com/business/archive/2016/06/the-trump-effect-and-cable-news/487472/

Wasko, J. (2014). The study of the political economy of the media in the twenty-first century. *International Journal of Media & Cultural Politics* 10(3), 251–279.

Wasko, J., Murdock, G., & Sousa, H. (2011). *The handbook of political economy of communications*. Malden, MA: Wiley-Blackwell.

Wojcik, N. (2017, May 3). Trump has been "rocket fuel" for NYT digital subscriptions, CEO says. CNBC. Retrieved from https://www.cnbc.com/2017/05/03/shares-of-new-york-times-surge-after-subscriber-growth.html

Yergin, D. (2008). *The prize: The epic quest for oil, money & power*. New York: Free Press.

3

A CASE OF UN-COVERAGE?

Deep media, Indigenous representation, and environmental issues

In his foreword to *Environmental Reporting in the 21st Century*, Bud Ward, editor of the *Yale Forum on Climate Change & the Media*, provides one way to think about contemporary environmental journalism (and journalism as a whole):

> Picture it this way: Think of a huge mansion. That's the house of journalism. All right, it's a severely dilapidated one at that. Environmentalism journalism, notwithstanding the enormity of the issues it embraces (think oceans, land, atmosphere, and beyond), constitutes one small room. No, make that one tiny closet of that mansion.
>
> *(2010, p. viii)*

Ward tells this story to bring to our attention that it is nearly impossible to fix the problems inherent in environmental reporting without first finding a solution for journalism as a whole. That is, why fix the closet when the roof is falling in?

The last chapter identified some of the structural, historical weaknesses that originate in the commercial logic of U.S. journalism, including close ties to government and corporations that manifest in a heavy reliance on public relations, official sources, and advertising – or what I have been terming *deep media*. While many of these problems were briefly discussed in the preceding chapter, here they are explored in depth regarding how they impact the reporting on environmental issues, including environmental justice. Boykoff and Boykoff (2007) observe that in order to interrogate media's communication about environmental issues, "research must critically scrutinize the firmly entrenched journalistic norms that profoundly shape the selection and composition of news" (p. 1191).[1] Towards that end, this section of the book progresses from an understanding of the broad problems facing journalism to explore how a commercial, largely corporate-owned U.S. media system covers a topic like the environment.

This then creates the foundation for the examination of how the press covered the events at Standing Rock in the following two chapters.

Seen from the perspective of *deep media*, the initial media blackout of #NODAPL is not only *not* unusual but is predictable, as many commercial news outlets have ignored Indigenous environmental justice struggles for decades if not for centuries (Moore & Lanthorn, 2017; Tallent, 2013; Klyde–Silverstein, 2012; Alia, 2010, 2004; Weston, 1996). Journalist Amy Goodman explains how the lack of coverage of the #NODAPL movement relates to the conventions of commercial journalism:

> It is astounding how little coverage the [Standing Rock Sioux Tribe] have gotten…. But this very much goes in lockstep with a lack of coverage of climate change. Add to it a group of people who are marginalised by the corporate media, Native Americans, and you have a combination that vanishes them.
>
> *(Al Jazeera, 2016)*

It is perhaps unsurprising, then, that commercial journalism initially ignored the struggles of the Tribe, for it continues to have a "blind spot" when it comes to the representation of both the environment and people of color. The media blackout of the protests thus reflects "two institutional biases in the US media: bias against environmental issues, as well as stories about Native Americans" (Al Jazeera, 2016). Seen from this perspective, #NODAPL's eventual prominence in the U.S. news media *is* unexpected and thus worthy of additional exploration.

In order to tease out the complexities related to the interconnection of Indigeneity, commercial journalism, and the environment, this chapter begins by making a case for continued exploration of the way in which traditional news media (increasingly known as "legacy media") cover a topic like the environment in the Internet age. A brief history of environmental reporting is provided as a way to lay the foundation to understand where we are today. This chapter defines environmental racism in relation to Indigeneity and then addresses how the structural limitations of conventional journalism impact coverage of environmental justice specifically. These discussions then lay the foundation to understand the media coverage of the Standing Rock resistance in 2016. As part of this foregrounding, I include examples of Indigenous environmental justice struggles in the U.S. and around the world.

Why press coverage of the environment matters

Years ago, Robert McChesney (2008) argued that "the greatest test" of the U.S. journalistic system was how it performed when the country went to war. The reason for his contention lay in the significant costs inherent to warfare: human life, relationships with other countries, intensive resource use, environmental impact, and more. While McChesney supports his contention well,

I would add that news coverage of the environment provides another, equally important litmus test. The increasing costs associated with large-scale environmental problems like climate change – including reconstruction efforts, the health of human lives and nonhuman animals, as well as the lives of future generations – means that there is much at stake when it comes to environmental reporting.

The importance of understanding how the news media cover environmental issues is underscored by the increasing attention being paid – by both the news media and the public – to global environmental disasters. As I write this in late summer 2017, another Category 5 Hurricane is roaring towards Puerto Rico after two others, Harvey and Irma, have devastated parts of Texas and Florida. The severity of these storms seems to have attracted significant public attention: scientists with the University of Wisconsin's *Space Science and Engineering Center* upgraded their categorization of Hurricane Harvey in particular as a "1,000-year" storm. A *Washington Post* article noted that the flooding from Harvey "is on an entirely different scale than what we've seen before in the United States," in the sense that "there is nothing in the historical record that rivals this" (Samenow, 2017). In addition to an especially strong hurricane season, the typically temperate and rainy Pacific Northwest had its driest and hottest summer on record in 2017, resulting in numerous forest fires that blanketed three states in ash and smoke for months while destroying numerous state parks and killing wildlife.

Many local and national news outlets made the connection between climate change and the severity of these powerful and destructive events clear: the *Seattle Times* and *The Atlantic* cited research claiming that climate change would increase both the number and the size of forest fires, while CNN and the *New York Times* rhetorically queried when the link between extreme weather events and climate change would be accepted more broadly. Many journalists have dared to draw a link between hurricanes Katrina, Sandy (also called "Superstorm Sandy"), and Harvey as evidence of catastrophic climate change. As one *Washington Post* article in the late summer noted, "the debate over climate and hurricanes is getting louder and louder" (Mooney, 2017).

It is difficult to overestimate the costs involved with environmental degradation. One can consider the actual costs of reconstruction after large-scale disasters: the recovery efforts in Texas cities after Hurricane Harvey is estimated to be in the tens of billions; to fight the 2017 forest fires in the state of Oregon alone cost the state at least $2 billion; and the reconstruction of the Gulf of Mexico after the British Petroleum oil spill cost more than $60 billion (according to BP). These costs, if substantial, are at least estimable through various calculations. More difficult to measure is the inherent value of a healthy forest or clean ocean. William Nordhaus (2013), economist and author of *Climate Casino*, notes that it is important to accurately gauge the value of ecosystems like oceans (or endangered animals, or land), precisely because it frames discussions around conservation in terms of economic value. From the standpoint of value, then, one can make appropriate claims for protection and mitigation. *Earth Economics*, a non-profit company based in Tacoma, Washington, systematically evaluates the economic value of natural resources. One of the natural resources they have evaluated is the large body of water known as Puget Sound. They write that

> Ecosystems within the Puget Sound Basin provide between $7.4 and $61.7 billion in benefits to people every year. If the 'natural capital' of the Puget Sound basin were treated as an economic asset, the asset value would be at least $243 billion to $2.1 trillion.
>
> *(Batker et al., 2008, p. 5)*[2]

It is undeniable that ecosystems have significant value (both estimable and incalculable), as human health is predicated upon a healthy environment. One of the best examples comes from the astronomically high levels of lead that recently have been discovered in the drinking water supply across the U.S. in recent years. National and local newspapers have reported on lead levels that were so high in Flint, Michigan drinking water that the water supply would fit the EPA's categorization of "toxic waste" (Ingraham, 2016). The national crisis of lead pollution in the water is especially hard on children's bodies, impacting their brains and central nervous systems. The *Houston Chronicle* reported that lead levels are 20 times higher than the EPA standards in elementary schools in southeast Houston, Texas. The school district in Tacoma, Washington is still discovering which schools' drinking water is unsafe: the water at one school, Reed Elementary, revealed lead levels "as high as 2,330 parts per billion. The Environmental Protection Agency's threshold for action by public water systems is 15 ppb" (Cafazzo, 2016). Intriguingly, the water testing results for Tacoma schools had come back a year prior, but only after a local paper – the *Tacoma News Tribune* – requested those reports did the school district decide to review the information and take action to inform parents. The fact that Reed Elementary has a majority Latino student population highlights another reason that media coverage is so important: environmental issues involve deep considerations of power and influence, especially when it comes to disadvantaged communities and historic neighborhoods of color. As a result, how the news media industry chooses to cover the environment, whether as a "problem" to be solved or simply as a broad topic, matters on numerous levels.

Does environmental journalism still matter in the age of the Internet?

The elephant in the room of journalism (and environmental reporting) is the Internet. Sachsman et al. (2010) highlight the seismic shifts in environment reporting due to the widespread adoption of new media:

> Google and other search engines have made this the golden age of information – and therefore the golden age of environmental information as well.

If one Googles "mountaintop removal" or "global warming," much more information is provided than can possibly be absorbed. Newspapers had been losing their dominance of the information business for many years with the development of new media from radio to television and now to the Internet. But throughout the 20th century, the best (though not the fastest) place to find news was in newspapers. In the 21st century, information has shifted dramatically to the Internet, and news – including environmental news – is moving there as well.

(p. 185)

Given that that information about environmental issues (and scientific information in general) is increasingly found on the Internet, is it still important to understand how the news media cover these issues? The answer is "yes," in large part because people still turn to legacy media for information they consider to be important. A recent Nielsen and Scarborough study revealed that 69 percent of Americans still read the news in numerous formats, including online, through apps, or in more traditional print[3] (Fletcher, 2016).

More specific to environmental news, the *Pew Research Center for Journalism and Media* notes that "Americans are most likely to get their science news from general news outlets and say the news media overall do a good job covering science" (Funk et al., 2017). The study reveals that for most Americans, social media play only a "modest role" when it comes to learning about issues related to science. In part this is because the majority of Americans seeking news still believe that the news media – despite the recent, widespread claims of their being "fake" – retain a significant amount of authenticity when it comes to the dissemination of scientific information. Of course, much of the reporting on the environment uses the lens of science, and so the idea of authenticity when it comes to news reports on scientific issues is extended to environmental ones as well. As Boykoff and Boykoff (2007, p. 1201) make clear, "Over time, the US news media have evolved into a powerful actor in the production, exchange, and dissemination of ideas within and between the science, policy and public spheres." Likely this is why members of an increasingly risk-averse society continue to trust the news media to inform them about impending environmental hazards (Lewis & Boyce, 2009).

Despite the panoply of information available on the Internet, then, "'legacy media' are still a primary means by which people orient themselves towards the world beyond their direct experience and, in particular, make sense of their own capacity for individual and collective agency" (Hackett et al., 2017, pp. 2–3). Armed with the evidence that the information-seeking public still turns to journalism, the case is made here that how traditional news media organizations report on environmental issues matters enough to study in depth. The next section thus turns to the history and background of what is often called "environment reporting" or environmental journalism.

Setting the stage: journalism, corporations, and environmental issues

As a news topic, the "environment" provides an audience draw for many reasons. Mike Dunne, a well-known environmental reporter for many decades, provides several reasons why:

> Environmental stories are great for [connection to the reader] because it's about the environment – it affects everybody. It's a political story. It's an economic story. It's a human interest story. It's a nature story. It covers everything."
>
> *(in Sachsman et al., 2010, p. 93)*

The news media began paying critical attention to the environment in earnest in the 1960s. Sachsman et al. (2010) provide a thorough treatment of the decades prior to this, noting that from the 1930s to the 1950s journalism tended to focus on engineering accomplishments, tying them to growing prosperity. In addition, businesses served as prominent underwriters for many TV and radio shows, creating a positive corporate image. At this time, pollution in particular was seen less as an urgent issue and more a mild cost of doing business:

> Throughout most of the Sixties, unless a river was on fire or a major city was in the midst of a weeklong smog alert, pollution was commonly accepted by the press and the general population as a fact of life.... Conservationists were thought of as eccentric woodsmen and environmentalists were considered unrealistic prophets of doom.
>
> *(Sachsman et al., 2010, p. 4)*[4]

One of the key reasons that the news media swept environmental pollution under the rug was that

> corporate public relations promoted this view, and skillfully kept the public satisfied. The press rarely heard the bad news of industry pollution but often received good-news releases concerning industry pollution controls and the many benefits offered to the community by local industry.
>
> *(p. 5)*[5]

Only corporations were producing the majority of press releases at this time, thereby shaping the agenda and creating the "fish-eye lens" to which Blyskal and Blyskal (1985) refer – a lens that kept their good deeds in focus while obscuring any practice that might be perceived as a problem.

The catalyst for the shift towards more critical environmental reporting is often attributed to Rachel Carson's *Silent Spring*, which, when published in 1962, was read by a public already "disgusted" by the widespread pollution of the water, land, and air that they attributed to corporations (Wyss, 2008). This growing distrust of business practices, including waste creation and disposal, was exacerbated by the release of data about corporate pollution through the creation of the *Freedom of Information Act* in 1966. Beder (1998) notes that "Between 1965 and 1970 environmental groups proliferated; environmental protection, especially pollution control, rose dramatically as a public priority in many countries. *Time* magazine labelled it a 'national obsession' in America" (p. 16). Sachsman et al. (2010) pinpoint the late 1960s as the starting period for environmental reporting due to prominent environmental disasters like the Union oil spill in the Santa Barbara Channel in 1969, where the news media depicted distraught young college students desperately trying to save oil-soaked seabirds. In fact, 1969 in particular seems to be a watershed year for reporting on the environment: the *New York Times* created an "environment beat," a practice other newspapers soon followed; *Time* magazine created an environment section; and *National Geographic* had a well-publicized article about the role of human activity in creating environmental problems (Sachsman et al., 2010).

If Carson's book helped to launch the environmental movement, then other events helped it rise to prominence, and the news media sat up and paid attention. The media industry wasn't the only one noticing, however: both the U.S. government and corporate America were alert to the changing tide of public opinion as well. In a three-year period alone (1969–1972), Congress passed four environmental laws (including the *Clean Air Act* and the *Clean Water Act*) and created (under Nixon) the *Environmental Protection Agency*. In addition, corporations were hit with many more regulations that favored the American citizen, including those that protected workers (OSHA) and consumers (including the Fair Credit Reporting Act). The raft of regulations created by the U.S. government during this time was, of course, not limited to the environment but represented sweeping changes in the way corporations had interacted with the government and the media for decades (Beder, 1998).

A positive shift for environmental reporting came in the late 1960s, when corporations ceased to be the only entity disseminating press releases; instead, the press started receiving information from environmental activist groups, universities, and various government agencies.[6] Sachsman et al. (2010) refer to this as the "environmental information explosion," one that was vehemently decried by industry: Richard Darrow, president of Hill & Knowlton public relations corporation, referred to this as the "Great Ecological Communications War – a war between conflicting public relations forces" (p. 6). As a result, the relationship that the business community had with the government, the press, and the public was changing, and corporations began to organize, mobilize, and strategize, in particular through the use of "environmental PR." While public environmental activism grew, the press became more keen to report critically on these issues, and the government began to provide more environmental regulations, the business industry was becoming more politically active in an attempt to shape the agenda. They created endowed chairs at universities, funded high schools and educational films, gave large amounts of money to politicians who supported corporate freedoms, created advertisements that touted their environmental records, and sponsored the arts (Beder, 1998). They also, of course, created new commercial relationships with the press.

The media are commercial: environmental reporting, "green" PR, and *spectacular environmentalisms*

Krovel (2012) notes that reporting on the environment tends to fall into a few distinct camps: "Small news items for those with a special interest in the environment; dramatic news on single issues; and poorly-camouflaged PR for business or industry" (p. 260).[7] There are many factors that help shape the public agenda when it comes to environmental issues, including the government, business, mass media, specialized environmental reporters, geographical factors, environmental activism, and academic institutions (Sachsman et al., 2010). And what shapes news coverage of environmental issues in particular? In part, it is the "traditional news values" discussed in the last chapter (including those related to professionalism and detachment), as well as the focus by the commercial press on "big events" that have "dramatic visual impact" (Sachsman et al., 2010, pp. 27–28). Both of these, of course, are related to the economic structure of commercial journalism, including the profit motive as well as the tensions that exist around funding of the news industry.

In Chapter Two, I noted how profit motivations and professionalism broadly impacts reporting and cited the *resource dependence model* as one of the best ways to understand the state of contemporary U.S. journalism. The model, which holds that an industry can be best understood by identifying who (or what) controls its resources (Grossberg et al., 2005; Turow, 1984; Aldrich & Pfeffer, 1976) is also incredibly useful to understand some of the central tensions and contradictions in environmental reporting. Lewis and Boyce (2009) observe that economist Victor Liebow's "clarion call to an environmentally wanton consumerism" has been embraced by the news media, undoubtedly because it coheres with the industry's advertising-as-revenue model (p. 6).

Reporting on the environment is thus particularly susceptible to corporate influence, whether through direct advertising, public relations, or the related practice of "native advertising" that has arisen in recent years. There is, in other words, a very close relationship between journalism and consumer culture, revealed especially by the three distinct ways that the press promotes commercial goods and services, including: when products are newsworthy (often in the case of controversy); during reviews of technology like cars and phones, of beauty products, or of cultural products like film); and through the public relations arm of the manufacturing industry (Schudson, 2013, p. 100). For Haluza-DeLay et al. (2009), the issue is simple: the news media reflect "corporate interests and promot[e] consumerist culture" (p. 13). A central problem is this: mainstream journalism's singular avoidance of critiques of consumption is antithetical to normative ideals of the press to act as a rigorous watchdog against the powerful. From the commercial news media landscape, then, what one might expect of coverage of the environment is that it would leans towards sensationalism and drama, rely upon the journalistic convention of "balance," be significantly influenced by public relations, and provide little or no context for complex issues. If this is true, in what ways does this occur, and in what type of cases? Through the use of prominent examples, the next few sections explore issues and tensions in environmental reporting.

"Green PR" and environmental reporting

In *Global Spin*, Sharon Beder (1998) provides perhaps one of the best treatments of how public relations has shaped environmental reporting. She deftly identifies the primary function of public relations, which is "to 'create news': to turn what are essentially advertisements into a form that fits news coverage and makes a journalist's job easier while at the same time promoting the interests of the client" (p. 113). There are distinct advantages for the corporate industry from using PR. First, despite its challenges and broad claims of "fake news," journalism continues to provide "credibility and legitimacy to what might otherwise be seen as self-serving publicity or advertising by giving it the appearance of being news delivered through... an 'independent' third party – the media" (p. 113). Second, prepackaged news from industry tends to crowd out investigative journalism that might shed light on questionable corporate practices. Finally, especially skillful PR can fundamentally shape a story, silencing alternative viewpoints and highlighting other aspects that are beneficial to a corporation or industry. Schudson (2013) agrees that PR can add more legitimacy than an overt advertisement, and adds another benefit it offers to corporations: it is relatively inexpensive, for firms typically do not pay for their material to be placed as a newspaper article.

While corporate interest in "environmental PR" has existed for decades (including when Monsanto tried to counter the environmental message of Rachel Carson's book *Silent Spring* in the early 1960s), the corporate PR machine really took off in the 1990s. Beder (1998) notes that

> By 1990 US firms were spending about $500 million a year on PR advice about how to green their image and deal with environmental opposition. By 1995 that figure had increased to a spend of about $1 billion per year on environmental PR activities. Today, most of the top PR firms include environmental PR as one of their specialties.
>
> *(p. 108)*

Citing those in the industry, Beder provides several reasons for why corporate "environmental PR" is on the rise. First, and perhaps most important, is that the majority of Americans express concern for the environment in various polls, and many consider themselves to be "environmentalists." Then there is the fact that information about corporations (eg., emissions and waste disposal) is now available to the public through the Freedom of Information Act. Finally, it is true that we live in an age of images, and that includes visual representations of environmental disasters and industrial accidents – ones that get the public's attention.

One example of a company using PR to its benefit during an environmental disaster was British Petroleum during its 2010 oil spill in the Gulf of Mexico. For 87 days, millions of Americans (and people from around the globe as well) watched in horror as over four million barrels of oil spilled into the gulf from an uncapped well 5,000 feet below the surface of the water. During a time of public outrage, BP began buying up search terms on Google. As a result, when an Internet user would search for "BP oil spill," one of the first links they would see on their results page was BP-sponsored information about its recovery efforts. In addition, as reported by cultural critic Jon Stewart on the *Daily Show*, BP intentionally released blurry video footage of the well leak instead of the HD version, which precluded scientists from accurately measuring how much oil was being spilled. One crucial part of PR is censorship (to stop negative information from reaching the public), and BP's efforts can be seen to fall within that category. Finally, BP spent more than $90 million – three times what it spent in years prior – on full-page color ads that ran in major dailies across the U.S. touting the company's cleanup efforts.

It is commonly estimated that at least half (and often more) of "news" stories originate in public relations (McChesney & Nichols, 2010; Alper & Jhally, 2002). The use of PR in environmental reporting falls in line with the commercial logic of the news industry discussed in Chapter Two: in an industry so reliant on advertiser dollars, the press release (and its cousin, the "video news release") is an easy and profitable way to create stories. The impact of public relations on the news coverage of environmental issues would be difficult if near impossible to overestimate, especially because the "best" PR is invisible – that is, if it is done right, the corporate footprint in any given news story is difficult to recognize (Alper & Jhally, 2002). Several scholars, however, have attempted to pull back the curtain to reveal how the PR industry works. Tim Holmes (2009) focuses on what he terms "denial industry" campaigns, especially by the tobacco and fossil fuel industries when it comes to urgent environmental issues that draw a lot of public attention. The now-infamous statement from a leaked tobacco industry memo in 1969 reads, "Doubt is our product since it is the best means of competing with the 'body of fact' that exists in the mind of the general public,"[8] providing a clear example of how the industry tried to protects its profits.

Of course, doubt campaigns are not limited to the tobacco industry: according to Holmes, "the campaign to cast doubt on climate change has been highly successful, and examples of the media's penetration by fossil fuel industry front-people abound" (p. 95). His research examined the "PR footprint" of BBC online, especially its relationship with ExxonMobil and its coverage of climate change. He found that, despite using sources that originated in the fossil-fuel industry, the BBC only acknowledged industry ties in less than a quarter of the articles. In addition, out of 88 articles, five referred only to the *Global Climate Coalition* without identifying it as "an energy industry lobbying group comprising companies such as Shell, BP, Amoco, and Exxon" (p. 96). Finally, he found "a large number of links to fossil-fuel industry funded organizations included… in sidebars or at the bottom of particular pages" as resources the BBC listed for further reading (p. 97). Holmes asks, somewhat rhetorically, if

> significant fossil fuel industry players have attempted to forestall or prevent action by muddying the waters on the facts of anthropogenic climate change, the question remains why such industry-friendly voices have been granted such a significant level of access to the media.
>
> *(p. 97)*

One of the reasons is, of course, professional journalism's practice of detachment and "balance," which is covered in more detail below; another reason, however, originates in journalism's economic structure – including its primary revenue source – that creates persistent vulnerabilities to corporate influence.

Advertising, Big Oil, and environmental coverage

As noted in the last chapter, the news media industry's economic structure is shifting. There has been a precipitous downturn of advertising revenue in the news media industry (Schauster et al., 2016; McChesney, 2014) due in part to the rise of big data and a new advertising strategy that tethers ads less to specific organizations (including newspapers) and more to individual users in what Couldry and Turow (2014) call "personalized advertising." In addition, individual subscriptions to national newspapers and news magazines are on the rise, which has – at least for national publications – shifted funding sources more towards subscribers. Although the economic landscape of the news media is transforming, this does not mean that advertising as a form of corporate influence is not important to consider. Large corporations still take out expensive ads in national dailies – British Petroleum spent roughly $5 million a week for full-page color ads in most major newspapers at the height of the Gulf of Mexico oil spill – and local businesses still turn to newspapers to promote their products to the local community.

In addition, while advertising often is seen only as a source of revenue, it undoubtedly remains "our most dominant and most ubiquitous cultural industry, with the few remaining non-commercial media…becoming conspicuous in their isolation" (Lewis & Boyce, 2009, p. 6). The trend towards increasing individual subscriber dollars can be interpreted as being due to the politics of "hate-watching Trump," but also due to the increasing awareness by corporations that audiences – whether online or in traditional print format – tend to dislike online ads. Part of the response by the news media industry to this has been to embrace "native advertising," where corporate-sponsored information (sometimes simply glorified press releases) masquerade as journalistic content. Schauster et al. (2016) attribute the precipitous rise in native advertising (outlets as prominent as the *New York Times* and *The Atlantic* have used it in recent years) to the decline of more traditional advertising support, noting that media organizations publish "sponsored content because its revenue combats the downturn in advertising revenue facing news organizations" (p. 1409). Similarly, Couldry and Turow (2014) note that recent "advertising dynamics are sweeping away the core 20th-century norm separating advertising from editorial matter in news" (p. 1718).[9]

Advertising (whether as overt ads or as "native advertising") is the "propaganda wing of consumer capitalism," whose "messages have colonized mediated communication and almost every public space" (Lewis & Boyce, 2009, p. 6). Schudson (2013) contends that "Even without public relations, commercial products are often the focus of news media attention. The media function as gatekeepers to cultural products" (p. 103). This blurring of the lines between the world of selling products and informing the public of important issues already creates friction when it comes to normative ideals regarding the role of a critical press, in large part because it is impossible to be a mouthpiece for corporate interest and a watchdog at the same time. And when it comes to reporting on the environment, this conflict of interest is made even more clear: while many environmental issues (e.g., climate change, pollution, shrinking biodiversity) could be at least partly mitigated by reducing consumption, the primary exhortation of the commercial news media is to consume. As Lewis and Boyce (2009) contend, "The looming environmental crisis brings the cultural meaning of advertising into much sharper focus" (p. 7).

One example of the close relationship between advertising, Big Oil, and journalism comes from an online, animated advertisement on the UK news site for *The Guardian* in late 2017 that featured an environmentally sensitive woman who met a giant "trying to be small." The giant – a bearded white male – was depicted trying to tiptoe through a field of flowers so as not to injure them. When they met, the giant knelt so as to be closer to the earth. The giant turned out to be an anthropomorphized version of energy giant Statoil, which drills offshore for oil and gas around the world. It also, according to this ad, invests in renewable energy. The intriguing part about Statoil's advertisements in *The Guardian* is that the newspaper published an article with a favorable treatment of Statoil a few months earlier in 2016. Titled "Green Really Is the New Black as Big Oil Gets a Taste for Renewables," the article noted several oil industry players, but gave the most article space to an uncritical discussion of Statoil's dedication to renewable energy and included direct quotes from both the company's chief executive and vice-president.

Another particularly illustrative example of the clear tension between advertising and environmental reporting comes from news media coverage of the 2010 oil spill in the Gulf of Mexico (Moore, 2011). As the oil was gushing uncontrollably into the gulf, the *Wall Street Journal* dedicated one page to the catastrophe that contained two articles: "BP Set to Try Risky Strategy to Contain Flow," and "Scientists to Back Dispersant Use, Despite Claims." Directly adjacent to these two articles was an advertisement (just as large as the articles) for Toyota's new "Star Safety System." On another day, the newspaper noted that the oil spill was now impacting the land and beaches, but on the same page, and double the size of the article, was an advertisement promoting the "JOY" of the new BMW X5 SUV. The front page of *USA Today* provided striking visuals of water and oil while asking in rather anxious large letters "Is it over?" Beneath the article, however, was an ad from a travel company exhorting readers to visit the Bahamas, showing a couple frolicking on a beach remarkably free of pollution. The tagline for the ad was – perhaps inadvertently – "It's Better Here." Another article, one from the *New York Times*, told the story of a turtle hurt by the spill: it was framed by a large ad (larger than the article) for American Airline's new nonstop flights to Madrid (Moore, 2011).

These examples revolve around a direct conflict addressed in this book, which is commercial journalism's ties to the oil industry and to oil consumption in general. As Lewis and Boyce (2009, p. 5) observe, "Few industries express this paean to [consumer culture] as enthusiastically as our media, information, and communication sector." News coverage of the 2010 oil spill in the Gulf of Mexico in particular reveals two key takeaways. The first is that commercial newspapers do not often directly challenge a consumer culture predicated on oil consumption. While news readers were confronted with the horrors of an uncontrolled oil spill, they simultaneously were encouraged to consume more of it – through flying, driving, and the purchase of endless types of merchandise. Out of hundreds of articles spanning the roughly 13 weeks of the spill, only one newspaper (*USA Today*) asked readers to critically examine their own role as oil consumers in environmental disasters like this one, while the other two newspapers restricted their discussions to the responsible parties to BP and the U.S. government (Moore, 2011). The second point is that readers are simultaneously hailed as citizens (who should care about the environment) *and* consumers (who should buy more to make themselves happy). The problem with this is that we cannot consume our way out of environmental problems, and if we approach these problems solely as consumers, the ecosystem will always lose: as the saying goes, "the only truly green product is the one you don't buy."

Newspapers' intimate ties to the business of selling create numerous problems for the ecosystem and the people who live in it. Lewis & Boyce (2009, p. 8) contend, that, even if

> the news media made climate change enemy number one, it would be in a commercial media culture in which the promotion of sustainability and a little less consumption appears, at best, joyless, and at worst, simply unimaginable. Or to put it another way, it is a media environment where constant exhortations to consume make it difficult for us to appreciate the fragility of the natural world and the environmental consequences of overconsumption.

The myriad ways in which the news media are influenced by corporate interest – whether through native advertising, behind-the-scenes public relations work, or direct advertisements – have a particular impact on the coverage of the environment. This next section details very specific ways in which commercial journalism's professionalism and profit drive impact the coverage of environmental problems. As Lewis and Boyce (2009, p. 9) have it, "The crux of the issue in how news media report on the environment revolves around two things: *how* climate change is represented and *how often* it is covered."

If it bleeds it leads: covering the "drama" of environmental events

Boykoff and Boykoff (2007) identify what they refer to as "first-order journalistic norms" in the news media: personalization, dramatization, and novelty: "we dub these 'first-order' journalistic norms because these factors are significant and baseline influences on both the selection of what is news and the content of news stories" (p. 1192). From the perspective of *deep media*, these journalistic conventions would be seen to be not only deeply woven into the newsmaking process but also related to both professionalism and the profit motive inherent in commercial journalism. The idea of *personalization* is very similar to Ingram's (2004) recognition of *individualization* in the media: namely, that environmental stories with a focus on personal triumph or tragedy displace meaningful discussions about the "big social, economic, or political picture" behind the issue or event (Boykoff & Boykoff, 2007, p. 1192). When stories are personalized, then, they omit broader considerations of context. This becomes especially significant when it comes to environmental issues, for oil spills (as one example) could be tied to the fossil-fuel scaffolding of our agricultural system as well as to consumption of goods and services.

Related to the *personalization* of a story, especially in relation to context, is *dramatization*. "Dramatized news tends to eschew significant and more comprehensive analysis of the enduring problems in favor of covering the spectacular machinations that sit at the surface of events" (Boykoff & Boykoff, 2007, p. 1192).[10] Alison Anderson (1997) observes that "environmental reporting tends to thrive on dramatic events involving 'goodies' and 'baddies'" (p. 7). She provides an example of dead seals washing up on the shores of the U.K. in the late 1980s, noting that the event had dramatic visuals due to the appeal of seals and the number of wounded and dead animals washing up on beaches. The old adage that "if it bleeds, it leads" prompts a question: does the focus on drama mean that most environmental stories are sensationalistic? The answer may be found in the distinction between episodic and thematic framing.

Episodic reporting on the environment: spectacular environmentalisms

Just as Boykoff and Boykoff (2007) identified 'first-order' journalistic norms, they also describe "second-order" conventions, including that news coverage is often episodic, contains an "authority-order bias," and is characterized by an effort to remain "balanced." Episodic reporting is considered here first. Schudson (2011) contends that most reporting can be characterized as "event-centered, action-centered, and person-centered," where "news focuses on visible events, often involving conflict or violent conflict" (p. 42). That is to say, many news stories are focused on one event: a single occurrence with a distinct beginning and ending, often involving a specific group of people responsible or involved. In the case of environmental issues, episodic reporting tends to focus on extreme weather events like hurricanes, calving glaciers, or a mass death of marine life.

The tendency towards event-centered reporting presents a challenge for commercial journalism, because while it has the benefit of creating relatively short, easily digestible articles for news audiences, it significantly undercuts the ability to cover complex processes with adequate context. That is, when the news is presented as a singular event, it not only encourages a superficial understanding of issues but also lacks the thematic framing needed to understand issues on a deeper level (Boykoff & Boykoff, 2007). This is especially important when it comes to environmental issues, because climate change is not an event but a process, one made immensely complex by: 1) the long time span that it encompasses; 2) the fact that its impacts are felt on numerous scales spanning from local to global; and 3) it manifests itself in so many different ways (e.g., floods, desertification, blizzards, forest fires, heat waves, species extinction, and ocean acidification). Thus, the record-breaking hurricane season and the significant calving of the Larsen C ice sheet in Antarctica in 2017 were typically covered as standalone events, but they could well be tied to climate change. The numerous deaths by starvation of orca calves in the Pacific Northwest's Puget Sound in 2017 were treated by local news media as a horrific but discrete event, although they could have been understood in relation to increasing fish consumption by humans and the effect of hydroelectric dams killing fish.

Because the news media are so focused on dramatic events, what the public gets most often are front-page stories about a "Snowpocalypse" (blizzard) and dramatic (and even conflict-driven) Pacific Northwest forest fires.[11] Often these news stories omit or marginalize the more complex causes that would help audiences to understand their context. Thus, "the environment appears to be an event-driven story…people seem to pay less attention to chronic environmental issues, where there are no acute events" (Sachsman et al., 2010, pp. 27–28). While one has to be careful in assuming what news audiences want, what we can know with certainty is that the commercial press tries to deliver relatively brief, attention-grabbing stories about people and events instead of deeper, more complex stories about more complex environmental processes.

Conflict-driven, dramatic, event-centered reporting on the environment has clear and direct parallels to *spectacular environmentalisms*, which Goodman et al. (2016, p. 678) define as "large-scale mediated spectacles about environmental problems." Examples of this type of environmentalism includes glossy magazine specials about environmental disasters, celebrity activity (in filmed marches and popular speeches), and perhaps even expensive, slick news productions with well-known reporters (e.g., "Our Year of Extremes" with NBC's Ann Curry in 2014). Being mediated is a key element of *spectacular environmentalisms*, in part because the media help mold these events into a commodity easily recognized in consumer culture. As Alison Anderson (2014) observes, "the combination of theatricality, drama, high profile celebrities, parody and satire communicated to global audiences via the web together with mainstream media, illustrates a new *spectacularisation* of the environment characteristic of power politics in late modern society" (p. 9).

The irony of these dramatic, mediated environmental statements is that they expend "vast amounts of CO2 to make vague gestures towards dealing with the climate crisis without critiquing corporate polluters or a model of economic growth that prioritizes increased production and profits…above the environment" (Goodman et al., 2016, p. 678). In so doing, this type of commodified environmental event has parallels to David Ingram's (2004) *mainstream environmentalism*, where the only viable solutions identified for environmental problems exist within a capitalist framework that reinforces consumption. It would be a stretch to say that environmental reporting performs this commodification of environmental problems in exactly this way, but it is undeniable that U.S. journalism operates within – and contributes to – the commercial media landscape critiqued by many scholars.

Seen from this perspective, much of the media coverage of the extremely destructive triad of hurricanes that hit the U.S. in late summer of 2017 would be considered not only episodic but also perhaps a type of *spectacular environmentalism*: heavily focused on visuals, with personalized human interest angles regarding the amount of human and animal suffering, but not addressing how it is we got here in the first place. While some outlets (notably *The Atlantic* and the *New York Times*) attempted to deepen their coverage by providing context – that is, was this powerful set of storms linked to a changing climate? – the majority treated the storms as discrete events that were best covered only in terms of how much damage was wrought, how many human lives were lost, and how much it would cost to rebuild. This type of coverage does not invite the public into deeper, thoughtful discussions about the role of humans in these changes, and what might be done as a result. That is, if the coverage of these types of events is reduced to more superficial, descriptive terms, news media audiences are not invited to question their role as consumer, nor are they challenged to take action as a citizen.

Professional journalism, detachment, and the coverage of environmental problems

The final critique of environmental reporting in this chapter involves a news media industry practice introduced in Chapter Two – professional journalism – and its reliance on notions of "balance." Boykoff (2011) notes that "Balance can... provide a 'validity check' for reporters who are on deadline and have neither the time nor scientific understanding to verify the legitimacy of various truth claims when covering complex issues in climate science and governance" (p. 108). While reporting both sides (the typical manifestation of balance) might initially sound commendable, it is one of the key *problems* associated with professional journalism's coverage of environmental issues. First, although hearing from both sides on any given issue sounds beneficial, there are often more than two sides. One exercise I do with my students revolves around them pretending to be reporters, whose focus is on presenting a "balanced" view of politics, covering the run-up to a U.S. presidential election. Inevitably, students will offer two perspectives from each of the two major parties: Republican and Democratic. But when I prod them by asking if those are the only possible perspectives, invariably they will come up with other possible sources they could have included in their story – the Libertarian, Green Party, Tea Party, and socialist perspectives. Thus, "balanced" reporting is often a significant reduction of possible perspectives by the news media, usually to two major ones, paradoxically (especially for the goal of being "balanced") giving an artificially narrow representation of the story.

Second, as Hansen (2010) observes, the practice of appearing objective often makes a journalist more reliant on official sources. McChesney (2008) agrees, contending that

> Professional journalist's core problem, and by no means its only problem, is that it devolved to rely heavily upon "official sources" as the basis of legitimate news. Official sources get to determine what professional journalists could be factually accurate about in the first place. It gets worse.... This reliance on official sources has always made professional journalism especially susceptible to well-funded corporate public relations, which could mask its interest behind a billion-dollar fig leaf of credentialed expertise. So it was that between 1995 and 2005 nary a single one of the 1,000 refereed academic research journal articles on climate change disputed the notion that something fundamental, dangerous, and influenced by humans was taking place. Yet our news media sources representing the interests of oil companies and other major polluters provided a significant official opposition to the notion that global warming was a problem or that pollution had anything to do with it. How significant? Over one-half of the 3,543 news articles in the popular press between 1991 and 2003 expressed doubt as to the existence...of global warming."
>
> *(p. 125)*

In discussing journalism's problem with official sources and the drive for impartiality, what McChesney makes very clear is that being "balanced" doesn't work well when reporting on climate change or any other environmental problems because it obscures the issue in favor of monied interests, including Big Oil. This leads to the third problem with "balanced" reporting of environmental issues, which is that it can make a problem like climate change seem like it is up for debate.

On his *Last Week Tonight* show in 2014, host John Oliver highlighted the problem with "unbiased" reporting on climate change. He noted that while over 97 percent of scientists were in agreement that climate change was real and caused by humans, the professional news' practice of presenting both sides made anthropogenic climate change seem completely debatable.[12] As Oreskes and Conway (2011) note, scientists have had a consensus about global warming and its causal link to human activity since the 1990s, "yet throughout this time period, the mass media presented global warming and its cause as a major debate" (p. 215). The consequence of journalism's balance on this environmental issue is that it "helped make it easy for our government to do nothing about global warming…Scientifically, global warming was an established fact. Politically, global warming was dead" (p. 215). In this sense, the practice of "balanced reporting" paradoxically creates its own bias, one that is "more conducive to the needs of media owners than to journalists or citizens" (McChesney, 2008, p. 125). The *Washington Post* article that drew attention to the "debate" on climate change during the 2017 hurricane season underscores the fact that, for commercial journalism, the debate over climate and hurricanes may indeed be getting "louder," but it still isn't settled.

A good summary of the biases inherent in environmental reporting, then, is to recognize that news:

> coverage is episodic and compartmentalized. It too infrequently connects the dots between, for example, the manifestations of climate change and its causes and consequences, or the rapid exploitation of fossil fuels and global warming. News routinely relies upon a narrow range of official sources… and overemphasizes disasters, rather than positive change agents and creative solutions. When solutions are addressed, there is too much focus on technology and individual 'green consumerism' rather than collective approaches and policy options. The overall editorial environment favours economic growth, consumption, and private sector business. There is little attention to who bears responsibility for climate change, and little critical analysis of capitalism or even the fossil fuel industry.
>
> *(Hackett et al., 2017, p. 4)*[13]

In terms of climate change, it is clear that the long-lasting public confusion about this issue "is largely due to systematic disinformation by the vested interests that would stand to lose from a lower-carbon economy, and from the hegemonic media's practice until recently of 'balancing' climate science with climate deniers" (Hackett et al., 2017, p. 3). With the recognition of how the structural weaknesses in journalism impacts reporting on the environment, we now explore commercial journalism's coverage of environmental justice issues.

The news covers environmental racism… or does it?
Profit, the environment, and "un" people

Hackett et al. (2017) observe that "too often, journalism overlooks marginalized people, vernacular languages, and traditional knowledge, and presents climate change as international news without local relevance" (p.3). This section focuses on commercial journalism's treatment of what Alia (2004) refers to as "un-people" – those considered by the media industry as insignificant, unimportant, and therefore undeserving of coverage.

Speaking directly to the issue of news media coverage of Indigenous groups, Mark Anthony Rolo writes:

> About this question of why the mainstream media continues to shut out the American Indian, it sure would be nice to say there was a time when this was not true. It would be nice to say there once was a "red" era in the history of the American press, an era when newspapers consistently published in-depth stories about the lives and struggles of Indian people, and that this expanse of coverage helped shape a more accurate image of Indians in the American public's mind. But, as stated, there is no "red" era of news coverage…. In fact, this poor record reveals the same kind of lame coverage the media has spent on all communities of color through the years. The American media has historically ignored these communities. The only thing more egregious than being shut out is to read inaccurate stories about your community that fuel stereotypes and even hatred.
>
> *(2000, p. 8)*

Anna Bean, Councilmember of the Puyallup Tribe in the state of Washington and a representative of *Puyallup Water Warriors*, provides an example of the difficulties Indigenous groups face in getting both the quantity and quality of coverage for their environmental concerns. *Puyallup Water Warriors* is a group that has been fighting the liquid natural gas plant proposed for the tide flats in Tacoma, Washington since 2016 and uses the hashtag #NOLNG253. There are many issues with the plant, but two central ones are that it would be constructed on the Tribe's ancestral lands and a large fossil fuel plant located on the water would have devastating impacts to people and wildlife if it failed. In conversations with me, Councilmember Bean noted the difficulties in getting adequate attention and fair coverage for their protests: "the media might show up early and make it look like no one's there, when really there were 400 people." She also expressed frustration regarding the way in which a local news outlet chose to edit the coverage of a member of the group whom they interviewed: "they edited out 45 minutes and in the end only had him talking about his grandparents," which she saw as taking his comments out of context.

In part, the lack of media attention originates in commercial news being unwilling to recognize or support existing Indigenous-produced media, despite the level of expertise and knowledge this type of media often offers (Alia, 2004, 1999). When the unwillingness of conventional journalism to cover Indigenous groups is paired with the sporadic, episodic, and "acontextual" coverage of environmental issues, it creates a significant gap in coverage that can have serious consequences – namely, that if the public remains unaware of a problem, it is unlikely to gain enough traction for meaningful change. This was especially true for the Standing Rock Sioux Tribe, who struggled to get adequate coverage from traditional news media until the *Democracy Now!* online video went viral in September 2016.

Environmental racism

Teresa Heinz (2005) defines environmental racism as "the placement of health-threatening structures such as landfills and factories near or in areas where the poor and ethnic minorities live" (p. 47). The importance of understanding how journalism covers environment racism is underscored by the clear and enduring racial and class-based disparity when it comes to access to clean land, air, and water. Bullard (2000), writing on environmental racism in the U.S., notes that "an abundance of documentation shows blacks, lower-income groups, and working class persons are subjected to a disproportionately large amount of pollution and other environmental stressors in their neighborhoods" (p. 1). In Louisiana's infamous "Cancer Alley," members of predominantly African-American communities struggle to sell their homes and move away from towns near major petrochemical industries. The impact of Hurricane Katrina on New Orleans in 2005 revealed that the poorest communities (predominantly African-American) lived at literally the lowest areas in the city – the ones most vulnerable to severe flooding (Davis, 2005). Similarly, the powerful Hurricane Harvey, which hit Texas in the summer of 2017, exposed the fact that majority-white neighborhoods in Houston had the political clout and money to afford greater structural protections against storms, while poorer communities of color were hit with more flooding and more polluted waters.[14]

The creation and continuance of what Steve Lerner (2010) refers to as "sacrifice zones" (low-income communities of color located near large, polluting industries) necessitates careful government monitoring and regulation to ensure healthy communities. Prior to the end of Barack Obama's presidency, the Environmental Protection Agency outlined its plan to help poorer communities near smelters and waste treatment plants address issues relating to air quality and lead poisoning (Milman, 2017). However, in early 2017, the Trump Administration proposed to cut the "environmental justice" office at the EPA – a move that would have "hit Black and [Latino] communities the hardest" (Milman, 2017). Congress rejected the majority of those cuts, thus temporarily saving the office, but the move drew attention to the important role that government can play in bridging the wide gap between white neighborhoods and communities of color. Despite the potential, often governmental action falls well short of protecting these communities.

One of the reasons may be that "environmental crime" (e.g., illegal dumping of industrial waste) is often difficult to identify, and even more difficult to attribute to a particular source (Jarrell, 2007). As one example, the city government of Seattle, Washington and the corporate giant Boeing have been engaged in a decades-long legal battle over who is responsible for the cleanup of polychlorinated biphenyls (PCBs) in the Duwamish River. The sticking point? Deciding who polluted the river in the first place. Decades-long debates like these mean that it is unlikely the river will be cleaned in the near future.

However difficult it may be at times to pinpoint the responsible party, it also is the case that local, state, and national governments tend to pay attention to highly mobilized communities with resources and political clout. The flip side of this coin, however, is that poorer neighborhoods with fewer resources often fall between the cracks and receive little or no help (Bullard & Wright, 2012; Davis, 2005; Bullard, 2000). It also is likely that the way in which environmental injustice cases are framed helps to perpetuate the problem: McGregor (2009) observes that the

> environmental justice literature...assumes a certain ideology about "environment" – one with a focus on how certain groups of people (especially those bearing labels such as "minority," "poor," "disadvantaged," or "Native") are impacted by environmental destruction, as if the environment were somehow separate from us.
>
> *(p. 27)*

In addition, some government policies may actually work against these communities. Writing on the U.S. government's *Nuclear Waste Policy Act*, Collins and Hall (1994) note, "One potential result of the government's policies is that people of color, having little political power, may bear the burden of an environmental problem that belongs to the entire nation" (p. 269).

Cases involving the environmental racism experienced by Indigenous groups are numerous and varied, including the impact of climate change and pollutants on Indigenous residents of the Arctic (Sakakibara, 2017; Davydov & Mikhailova, 2011); the long-lasting effects of uranium mining on the Navajo Nation (Pasternak, 2010); coal mining on the Salish Sea (LaDuke, 2017b); dangerously low water quality experienced by the First Nation Anishnaabe tribe (McGregor, 2009); and removal of natural resources from unceded tribal lands (Klein, 2014; Lawson, 2009; Erdoes, 1982). In another case, the Lower Elwha Klallam Tribe in Washington State had much of its land eradicated in the early 20th century by the erection of a hydroelectric dam (Moore & Lanthorn, 2017; Crane, 2011). That case clearly highlights that although one can conceive of cases involving environmental racism strictly in terms of pollution, to have such a limited focus would ignore the unique environmental concerns facing Indigenous groups. Vickery and Hunter (2016) make this point directly: "common measures reflecting ...distance to hazardous facilities do not capture the complexity of Native

American connections to landscape" (p. 3). The reason for this is that there are many Tribes who consider their identity to be intimately tied to the land, a connection that Harris and Harper (2011) refer to as an *ethno-habitat*.

Like the Standing Rock Sioux Tribe and their struggle against an oil pipeline, Indigenous groups have often found themselves, almost always unwillingly, at the epicenter of fossil fuel projects. Noted above, the Puyallup Tribe is fighting a large, liquified natural gas facility proposed for the water's edge of their Salish Sea on their ancestral land. Nearby, the Lummi Nation was successful in warding off a proposed coal terminal that would have endangered their fishing rights by treaty. In other areas of the U.S., the Nez Perce, or *Nimi'ipuu*,[15] have stopped large oil trucks passing through its reservation on the way to oilfields, and the Northern Cheyenne are successfully resisting coal development in areas of Montana (Klein, 2014).

In addition, many Indigenous groups have been inextricably linked to the U.S. nuclear energy project in terms of both production as well as waste storage. Collins and Hall (1994) observe that

> The United States, like other nations, exploits the uranium resources found on lands retained by aboriginal communities. The path of nuclear mining and milling in the United States has led repeatedly across Indian country, leaving a legacy of nuclear waste and contamination. Today, in its quest to rid the nation of stockpiles of highly radioactive waste from civilian nuclear power plants, the federal government once again turns to Indian country.
>
> *(pp. 268–269)*

Given all these examples, McGregor (2009) appropriately defines environmental justice as it relates to Indigeneity through the lens of power: "From the perspective of the world view within which I am embedded, environmental justice is most certainly about power relationships among people and between people and various institutions of colonization" (p. 27).

The role of the news media in reporting on environmental justice

That the government often falls short in protecting poor and working-class people of color from land, air, and water pollution reinforces the need for the news media to act as watchdog on behalf of the public. Journalism can draw attention to previously unknown environmental justice issues, enabling public discussion and, perhaps, action. In the U.S. there have been a few prominent cases of investigative journalism that revealed environmental justice to the public. *Carnegie-Knight News 21*, an investigative reporting initiative that involves students from multiple universities, recently released a report run by students titled "Troubled Waters," which found that as many as 63 million people in the U.S. (including those from 'disadvantaged' communities) had been exposed to unsafe drinking water within the last decade. Al Jazeera's Massoud Hayoun, writing for the news outlet's American branch,[16] did a three-part series on a multi-billion dollar methanol plant funded by Chinese investors, including a report on how the plant would negatively impact the nearby historic African-American community. In 2012, *USA Today* created a multi-part, interactive series titled "Ghost Factories," which won a science award from *National Academies* for the exploration of old lead smelting sites and their impacts on local communities. National Public Radio published a four-part series titled "Poisoned Places: Toxic Air, Neglected Communities" in 2011 about air pollution, complete with an interactive map where readers could identify pollution in their own neighborhood.

Globally, there are other examples of journalists and news agencies working to shed light on significant environmental issues. Mike Anane, a Ghanaian journalist and former editor of *The Triumph* newspaper, has revealed how the practice of dumping European and American e-waste in his country has devastated formerly pristine wetlands, including the district of Agbogbloshie – the largest e-waste site in the world. Tran Thi Thuy Binh, a journalist in Vietnam, covers environmental issues like biodiversity protection and climate change and community-based conservation models. In June of 2018 she published a three-part series investigating the connection between climate change, poverty, migration, and women in her country.[17] She also has written about environmental topics such as air pollution near metal smelting sites in Vietnam[18] and collaborated on "A Coastal Green Shield of Community," a documentary about planting mangrove trees to protect her country's coastline. In another example of global work, Stoddart (2007) analyzed the *Vancouver Sun*'s coverage of environmental justice, finding that the newspaper simplified environmental policy debate into a conflict between the government, the forest industry, and environmental groups, marginalizing the perspective and voices of First Nations groups.

These examples, although laudable, are conspicuous in their isolation, and environmental justice remains an issue largely ignored by mainstream news media. In the U.S., environmental justice cases most often receive scant consideration in commercial journalism (Klyde-Silverstein, 2012; Heinz, 2005; Rosen, 1994). When journalism does pay attention to environmental justice issues, the coverage can be sensationalized, and often contains stereotypes of the people involved or minimization of the environmental problem (Loew, 2011; Heinz, 2005; Alia, 2004). In fact, environmental problems tend to receive sparse news attention unless they occur in a higher-income, white neighborhood with more resources and political power (Heinz, 2005). In the Global North outside the U.S., research reveals that "Canadian newspapers subtly downplay environmental threats to marginalized communities" and tend to focus on "economic and socio-political threats (e.g., social unity) at the expense of environmental threats" (Deacon et al., 2015, p. 429). According to Haluza-DeLay et al. (2009), the mainstream media are complicit in what they term the "invisibilization" of the environmental experiences, knowledge, and needs of Aboriginal peoples in Canada: "for the most part, the voices of Aboriginal peoples have been ignored, dismissed, or overridden since contact" by the Canadian government, and they see the media as being complicit in this (p. 3). Acknowledging how the conventional news media have performed in reporting on environmental justice in the past, the next chapter turns to the focal point of this book: exploring how the news media covered the events at Standing Rock.

There is no 'post' in post-colonial: *deep media* and environmental justice

In her essay "Finding a New Voice" about media representations of Indigenous groups, Patty Loew observes that

> If art imitates life, the world has long gazed upon a surrealistic portrait of Indigenous people... . However, these are not true Native portraits. These are the representations of white colonists, military leaders, government workers, and missionaries whose lives intersected with Native Americans.
>
> *(2011, p. 3)*

Loew notes the specific trend in mainstream news media to sensationalize stories about Native Americans (while ignoring crucial context) in a way that can trivialize and misrepresent Indigenous people while minimizing their struggles.

Chapter Two laid the groundwork for understanding contemporary, commercial journalism through the lens of *deep media* – those deeply ingrained, historical conventions woven into the newsmaking process that are structured by professionalism, profit motive, and close connections to the corporate world, including the oil industry. This chapter has revealed how the structural vulnerabilities seen in *deep media* impact coverage of environmental problems broadly, and environmental justice specifically. Framing the discussion through the lens of *deep media* keeps the focus on how the structures woven into media influence this coverage:

> While it is right to hold the news media to account for the way they report [environmental issues like] climate change, the battleground is much larger than this. It encompasses the whole deregulatory, commercial thrust of media and telecommunications policy...that has stressed product proliferation over content and the increasing incursions of advertising into all forms of communications.
>
> *(Lewis & Boyce, 2009, p. 9)*

When it comes to reporting on environmental justice, for researchers like Haluza-DeLay et al. (2009), the line distinguishing the corporate world and the media industry is fluid and permeable: "the marginalization of First Nations is an effect of state and corporate (including media and education) strategies for appropriating land and exploiting or diverting natural resources from land use and occupancy to extraction and sale… ." This interlocking of the government, the corporate world, and the media industry means that, "from an aboriginal perspective, there is no 'post' in post-colonial" (p. 16).

Understanding that commercial journalism has marginalized, ignored, or sensationalized environmental problems, including those faced by people of color, lays the foundation for analysis of mainstream news coverage of the Standing Rock Sioux Tribe's #NODAPL movement in 2016. What, if anything, has changed? What remains the same? This is the focus of the next two chapters.

Notes

1 Although Boykoff and Boykoff (2007) are speaking directly to "climate science communication" (that is, about mediated representations of climate change), their observations can be extrapolated to communication about all types of environmental issues.

2 As another example, Earth Economics estimated the value of a local park. They examined first the seemingly obvious: how many trees are there and how much carbon dioxide do they absorb? However, then they deepened their analysis to examine how much the city would save in medical costs (because the park contains a running track as well as the mental health benefits from getting outside and walking).

3 The original Nielsen report is available here: http://www.nielsen.com/us/en/insights/news/2016/newspapers-deliver-across-the-ages.html?cid=socSprinklr-Nielsen

4 Sachsman et al. (2010) cite the dissertation of Sachsman (1973) for the original version of this.

5 Sachsman et al. (2010) provide the example of the company International Harvester, which burned coal. Their proactive communication with the community (they promised to take care of the problem of the coal dust that was blanketing the nearby residential communities) meant that citizens waited patiently for years without complaining, and "the first newspaper story ever carried on the issue was headlined, 'IH Spends $71,900 to Be a Good Neighbor.'" (p. 5).

6 Although Sachsman et al. note that the government is the "dominant source" for environmental news, one must ponder how the government organizations like the Environmental Protection Agency, National Park Service, and National Aeronautics and Space Administration will respond to the Trump administration's seeming desire to mute environmental information (such as records of animal abuse, scientific evidence for climate change, etc.).

7 Krovel cites the origin of this observation as Morck (1996) in Jan Johnsen's *Milijo Medier Og Skole*, Rapport Publishers (in Norwegian).

8 The original document containing this statement can be found at https://www.industrydocumentslibrary.ucsf.edu/tobacco/docs/#id=xqkd0134

9 More specifically, Couldry and Turow (2014) note that there are three new dynamics that are reshaping the media landscape: "the growing prevalence of personalized advertising based on continuous data mining, leading to pervasive pressures to personalize content in response" (p. 1718).

10 Boykoff and Boykoff (2007) cite Wilkins and Patterson (1987) for this.

11 In the case of the fires that devastated the Eagle Creek forest in Oregon in 2017, the conflict centered around a group of teenagers who were lighting fireworks and throwing them into the woods. When told by hikers that there was a drought and that forest fires were likely, they laughed and threw another firecracker into the brush. Minutes later, flames engulfed the area and burned for weeks. The local media reported more on the interpersonal conflict itself than on why the forests were so dry (climate change impacting the Pacific Northwest).

12 In addition, many of the scientists in the minority regarding human-caused climate change were found to have some ties to – or were funded by – the oil and gas industry. One example comes from Goldenberg (2015), who reports that scientist Willie Soon's "work was funded almost entirely by the energy industry, receiving more than $1.2m from companies, lobby groups and oil billionaires over more than a decade."

13 Hackett et al. (2017) cite interviews they conducted with climate change communicators as part of the *Climate Justice Project* in Vancouver, British Columbia.

14 In addition, Bullard (2000) observes that because Houston has no zoning, the "city's landscape has been shaped by haphazard and irrational land-use planning" (p. 40), a situation that has given rise to more environmental protection for those communities with more resources and political sway.

15 Nez Perce is a popular but inexact exonym of French origin for the Tribe meaning "pierced nose," but the Tribe refers to itself as "Nimi'ipuu."

16 The American branch of Al Jazeera closed in 2016.

17 Tran Thi Thy Binh's work on this can be found at HanoiTV: http://hanoitv.vn/mat-co-hoi-chon-viec-do-sat-lo-bo-bien-d93264.html

18 Her work on the air pollution caused by metal smelting is available at http://hanoitv.vn/nhung-nguoi-khong-chon-song-sach-phan-1-d20280.html

References

Aldrich, H. E., & Pfeffer, J. (1976). Environments of organizations. *Annual Review of Sociology, 2,* 79–105.

Al Jazeera. (2016, December 11). Dakota access pipeline: Behind the 'media blackout'. Al Jazeera. Retrieved from http://www.aljazeera.com/programmes/listeningpost/2016/12/dakota-access-pipeline-media-blackout-161210071122040.html

Alia, V. (1999). *Un/covering the north news, media, and aboriginal people.* Vancouver, BC: UBC Press.

Alia, V. (2004). *Media ethics and social change.* New York: Routledge.

Alia, V. (2010). *The new media nation: Indigenous peoples and global communication.* New York: Berghahn Books.

Alper, L., & Jhally, S. (Directors). (2002). *Toxic sludge is good for you: The public relations industry unspun.* [Video/DVD] Northampton, MA: Media Education Foundation.

Anderson, A. (1997). *Media, culture and the environment.* London: Routledge.

Anderson, A. (2014). *Media, environment and the network society.* New York: Palgrave Macmillan.

Batker, D., Swedeen, P., & Costanza, R. (2008). A new view of the Puget Sound economy. *Earth Economics.* Retrieved from https://www.floods.org/ace-files/documentlibrary/committees/A_New_View_of_the_Puget_Sound_Economy.pdf

Beder, S. (1998). *Global spin: The corporate assault on environmentalism.* White River Junction, VT: Chelsea Green Pub.

Blyskal, M. H., & Blyskal, J. (1985). *PR: How the public relations industry writes the news.* New York: W. Morrow.

Boykoff, M. T. (2011). *Who speaks for the climate? Making sense of media reporting on climate change.* UK: Cambridge University Press.

Boykoff, M. T., & Boykoff, J. M. (2007). Climate change and journalistic norms: A case-study of US mass-media coverage. *Geoforum, 38*(6), 1190–1204. 10.1016/j.geoforum.2007.01.008

Bullard, R. (2000). *Dumping in Dixie: Race, class, and environmental quality* (3rd ed.). Boulder, CO: Westview Press.

Bullard, R., & Wright, B. (2012). In B. Wright (Ed.), *The wrong complexion for protection: How the government response to disaster endangers African American communities.* New York: New York University Press.

Cafazzo, D. (2016, April 26). High lead levels found in drinking water in 4 more Tacoma schools. *The News Tribune.* Retrieved from http://www.thenewstribune.com/news/local/education/article73952142.html

Collins, N. B., & Hall, A. (1994). Nuclear waste in Indian country: A paradoxical trade. *Law & Inequality: A Journal of Theory and Practice, 12*(2), 267.

Couldry, N., & Turow, J. (2014). Advertising, big data, and the clearance of the public realm: Marketers' new approaches to the content subsidy. *International Journal of Communication, 8*, 1710–1726.

Crane, J. (2011). *Finding the river: An environmental history of the Elwha.* Corvallis, OR: Oregon State University Press.

Davis, M. (2005, September). The predators of New Orleans. *Le Monde Diplomatique.* Retrieved from http://kit.mondediplo.com/spip.php?article4263

Davydov, A., & Mikhailova, G. (2011). Climate change and consequences in the arctic: Perception of climate change by the Nenets people of Vaigach island. *Global Health Action, 4*(1), 10.3402/gha.v4i0.8436.

Deacon, L., Baxter, J., & Buzzelli, M. (2015). Environmental justice: An exploratory snapshot through the lens of Canada's mainstream news media. *Canadian Geographer / Le Géographe Canadien, 59*(4), 419–432. 10.1111/cag.12223

Erdoes, R. (1982). *Native Americans, the Sioux.* New York: Sterling Pub. Co.

Fletcher, P. (2016, December 26). Good news for newspapers: 69% of U.S. population still reading. *Forbes.* Retrieved from https://www.forbes.com/sites/paulfletcher/2016/12/26/good-news-for-newspapers-69-of-u-s-population-still-reading/#c38e0c6723c8

Funk, C., Gottfriend, J., & Mitchell, A. (2017). A majority of Americans rely on general outlets for science news but more say specialty sources get the facts right about science. *Pew Research Center of Journalism and Media.* Retrieved from http://www.journalism.org/2017/09/20/science-news-and-information-today/

Goodman, M. K., Littler, J., Brockington, D., & Boykoff, M. (2016). Spectacular environmentalisms: Media, knowledge and the framing of ecological politics. *Environmental Communication, 10*(6), 677–688. 10.1080/17524032.2016.1219489.

Grossberg, L., Wartella, E., Whitney, C., & Wise, J. (2005). *Mediamaking: Mass media in a popular culture* (2nd ed.). Thousand Oaks, CA: Sage Publications.

Hackett, R. A., Forde, S., Gunster, S., Foxwell-Norton, K., Hackett, R. A., Forde, S., Foxwell-Norton, K. (2017). *Journalism and climate crisis: Public engagement, media alternatives.* New York: Routledge.

Haluza-DeLay, R., O'Riley, P., Cole, P., & Agyeman, J. (2009). Introduction. Speaking for ourselves, speaking together: Environmental justice in Canada. In J. Agyeman, P. Cole, R. Haluza-DeLay, & P. O'Riley (Eds.), *Speaking for ourselves: Environmental justice in Canada* (pp. 1–26). Vancouver, BC: UBC Press.

Hansen, A. (2010). *Environment, media and communication*. London: Routledge.

Harris, S., & Harper, B. (2011). A method for tribal environmental justice analysis. *Environmental Justice, 4*(4), 231–237.

Heinz, T. L. (2005). From civil rights to environmental rights: Constructions of race, community, and identity in three African American newspapers' coverage of the environmental justice movement. *Journal of Communication Inquiry, 29*(1), 47–65. 10.1177/0196859904269996

Holmes, T. (2009). Balancing acts: PR, "impartiality," and power in mass media coverage of climate change. In T. Boyce & J. Lewis (Eds.), *Climate change and the media* (pp. 92–101). New York: Peter Lang.

Ingraham, C. (2016, January 16). This is how toxic flint's water really is. *Washington Post*. Retrieved from https://www.washingtonpost.com/news/wonk/wp/2016/01/15/this-is-how-toxic-flints-water-really-is/?utm_term=.9f9c4576d9db

Ingram, D. (2004). *Green screen: Environmentalism and Hollywood cinema*. Exeter, UK: University of Exeter Press.

Jarrell, M. (2007). *Environmental crime and the media: News coverage of petroleum refining industry violations*. New York: LFB Scholarly Pub.

Klein, N. (2014). *This changes everything: Capitalism vs. the climate*. New York: Simon & Schuster.

Klyde-Silverstein, L. (2012). The 'Fighting whites' phenomenon: An interpretive analysis of media coverage of an American Indian mascot issue. In M. Carstarphen & J. Sanchez (Eds.), *American Indians and the Mass Media* (pp. 113–127). Norman, OK: University of Oklahoma Press.

Krøvel, R. (2012). Setting the agenda on environmental news in Norway: NGOs and newspapers. *Journalism Studies, 13*(2), 259–276. 10.1080/1461670X.20311.646402

LaDuke, W. (2017b). *The Winona LaDuke chronicles: Stories from the front lines in the battle for environmental justice*. Nova Scotia: Fernwood Publishing.

Lawson, M. L. (2009). *Dammed Indians revisited: The continuing history of the Pick-Sloan Plan and the Missouri River Sioux*. Pierre, SD: South Dakota State Historical Society Press.

Lerner, S. (2010). *Sacrifice zones: The front lines of toxic chemical exposure in the United States*. Cambridge, MA: MIT Press.

Lewis, J., & Boyce, T. (2009). Climate change and the media: The scale of the challenge. In J. Lewis & T. Boyce (Eds.), *Climate Change and the Media* (pp. 3–16). New York: Peter Lang.

Loew, P. (2011). Finding a new voice – foundations for American Indian media. In M. Carstarphen & J. Sanchez (Eds.), *American Indians and the mass media* (pp. 3–6). Norman, OK: University of Oklahoma Press.

McChesney, R. (2014). The struggle for democratic media: Lessons from the north and from the left. In C. Martens, E. Vivares, & R. McChesney (Eds.), *The international political economy of communication: Media and power in South America* (pp. 11–30). London: Palgrave Macmillan.

McChesney, R. W. (2008). *The political economy of media: Enduring issues, emerging dilemmas*. New York: Monthly Review Press.

McChesney, R., & Nichols, J. (2010). *The death and life of American journalism*. New York: The Nation Books.

McGregor, D. (2009). Honoring our relations: An Anishnaabe perspective on environmental justice. In J. Agyeman (Ed.), *Speaking for ourselves: Environmental justice in Canada* (pp. 27–41). Vancouver, BC: UBC Press.

Milman, O. (2017, March 3). 'Just racist': EPA cuts will hit black and Hispanic communities the hardest. *The Guardian*. Retrieved from https://www.theguardian.com/environ ment/2017/mar/03/epa-environment-budget-cuts-pollution-justice-office

Mooney, C. (2017, Aug 30). Katrina, Sandy, Harvey. The debate over climate and hurricanes is getting louder and louder. *Washington Post*. Retrieved from https://www. washingtonpost.com/news/energy-environment/wp/2017/08/30/katrina-sandy-harvey-the-debate-over-climate-and-hurricanes-is-getting-louder-and-louder/? utm_term=.3b26ca419e89

Moore, E. (2011). Spill baby spill: Exploring commercial journalism coverage of the 2010 gulf oil spill. Paper presented at the *National Communication Association*, New Orleans.

Moore, E. E., & Lanthorn, K. R. (2017). Framing disaster. *Journal of Communication Inquiry, 41*(3), 227–249. 10.1177/0196859917706348.

Nordhaus, W. D. (2013). *The climate casino: Risk, uncertainty, and economics for a warming world*. New Haven, CT: Yale University Press.

Oreskes, N., & Conway, E. (2011). *Merchants of doubt: How a handful of scientists obscured the truth on issues from tobacco smoke to global warming*. New York: Bloomsbury Press.

Pasternak, J. (2010). *Yellow dirt: An American story of a poisoned land and a people betrayed*. New York: Free Press.

Rolo, M. A. (2000). Introduction. In M. A. Rolo (Ed.), *The American Indian and the media* (2nd ed., pp. 8–13). St Paul, Minnesota.

Rosen, R. (1994). Who gets polluted – the movement for environmental justice. *Dissent, 41*(2), 223–230.

Sachsman, D. B., Simon, J., & Valenti, J. M. (2010). *Environment reporters in the 21st century*. New Brunswick, NJ: Transaction Publ.

Sachsman, D. B. (1973). *Public relations influence on environmental coverage (in the San Francisco bay area)*. Dissertation, Stanford University. Retrieved from https://search.proquest. com/docview/302677921

Sakakibara, C. (2017). People of the whales: Climate change and cultural resilience among iñupiat of arctic alaska. *Geographical Review, 107*(1), 159–184. 10.1111/j.1931-0846.2016.12219.x

Samenow, J. (2017, August 31). Harvey is a 1,000-year flood event unprecedented in scale. *The Washington Post*. Retrieved from https://www.washingtonpost.com/news/capital-weather-gang/wp/2017/08/31/harvey-is-a-1000-year-flood-event-unprecedented-in-scale/?utm_term=.71b413fb7488

Schauster, E. E., Ferrucci, P., Neill, M. S., Wojdynski, B. W., & Golan, G. J. (2016). Native advertising is the new journalism. *American Behavioral Scientist, 60*(12), 1408–1424. 10.1177/0002764216660135

Schudson, M. (2011). *The sociology of news*. (2nd ed.) New York: W.W. Norton & Company.

Schudson, M. (2013). *Advertising, the uneasy persuasion: Its dubious impact on American society*. London: Routledge.

Stoddart, M. C. J. (2007). 'British Columbia is open for business': Environmental justice and working forest news in the Vancouver Sun. *Local Environment, 12*(6), 663–674. 10.1080/13549830701664113

Tallent, R. (2013, May 1). Don't misrepresent Native Americans. *The Quill, 101*, 26.

Turow, J. (1984). *Media industries: The production of news and entertainment*. New York: Longman.

Vickery, J., & Hunter, L. M. (2016). Native Americans: Where in environmental justice research? *Society & Natural Resources, 29*(1), 36–52. 10.1080/08941920.2015.1045644

Ward, B. (2010). Foreword. In J. Simon, J. Myer Valenti, & D. B. Sachsman (Eds.), *Environment reporters in the 21st century* (pp. ix). London: Transaction Publishers.

Weston, M. A. (1996). *Native Americans in the news: Images of Indians in the twentieth century press*. Westport, CT: Greenwood Press.

Wyss, B. (2008). *Covering the environment: How journalists work the green beat*. New York: Routledge.

4

FRAMING INJUSTICE

U.S. media coverage of the Standing Rock movement

In struggles for environmental justice, whose voices are recognized and given consideration by the news media? And how are voices, when they are acknowledged, framed for news audiences? As the preceding chapters have made clear, the mainstream news media contain deep flaws and vulnerabilities that impact coverage of myriad issues, including environmental problems like pollution and climate change. When considering environmental justice, it becomes evident that many of the charges that have been leveled at commercial news coverage of the environment – that stories are infrequent, episodic, often sensationalized, focused on objectivity and "balance," and lack context – can also be applied to journalism's treatment of Indigenous groups. As a result, those groups seeking timely, consistent, and fair coverage for their environmental justice concerns need to pass the first hurdle: namely, the news industry's tendency to marginalize or ignore *both* the environment and Indigenous people.

While the first three chapters of this book provided systematic critiques of the U.S. commercial news media system in relation to the coverage of environmental issues, this section deepens the critical exploration through a framing analysis of news coverage of the construction of the Dakota Access Pipeline and the resistance movement that grew around it (often referred to as #NODAPL on social media). The importance of studying this is underscored by the recognition of the power of mediated discourse, which "is not only vital in terms of framing social issues and problems for the attentive public, but it is also a place of ideological and ideational struggle for various social movements, state actors, and institutions" (Boykoff, 2006, p. 227). When it comes to coverage of environmental justice, then, it is clear that these issues involve "deep questions about value," especially in regard to who merits attention, consideration, and respect (Moore & Lanthorn, 2017, p. 228).

The central aim of this chapter and the next one is to discover how various news sources in the U.S. and Canada covered the environmental struggles faced by the Standing Rock Sioux Tribe. That is, how did newspapers frame the struggle over the pipeline? What factors influenced this coverage? How do the results speak to issues of power when evaluating the complex delineations of environmental justice cases? When considering the contours of power and the construction of meaning, it is important to recognize that the struggle by the Tribe is "tied to broader issues of social justice and the political structures that have long sustained marginalization of low income and racial/ethnic minority groups" (Deacon et al., 2015, p. 420). Frames, then, are one way in which deeper values and meanings can be revealed.

The utility of framing analysis for understanding the complexities of news media coverage of #NODAPL thus comes from recognition that frames provide an effective "way to describe the power of a communicating text" (Entman, 1993, p. 51). Popularly defined by Gitlin (1980) as "principles of selection, emphasis, and presentation" that reveal "little tacit theories about what exists, what happens, and what matters" (p. 6), frames have been the subject of a great deal of scholarly debate – including how it should be done (Matthes & Kohring, 2008), whether it constitutes both method and theory (Scheufele, 1999), and the replicability of analysis (David et al., 2011). The intent here is not to wade into specific deliberations and controversies about framing but instead to first identify why it is an appropriate framework to answer the questions posed in this research and make clear how it is defined and applied in this particular context. Before a discussion of framing as an interpretive framework for this research, it is important to recognize existing framing scholarship on two subjects: Indigeneity and environmental justice.

News framing of Indigenous groups

When evaluating the #NODAPL movement, it is impossible to avoid consideration of how Indigeneity and environmental justice have been covered by the news media, especially because common depictions of Indigenous groups can directly influence coverage of the environmental issues faced by these groups. Noted in the previous chapter, the mainstream press has tended to ignore Native Americans, treating them as the "un people" that Alia (2004) describes – those people and groups of people who are seen as unimportant and therefore unworthy of news coverage. Larson (2006) notes that "topics important to Native Americans (such as land claims) are not typically covered in the mainstream press, which considers the audience for this news small and unimportant" (p. 108).

When Indigenous groups *do* receive coverage, how are they framed for news audiences? News stories often are sensationalized, tend to reinforce stereotypes (many involving poverty, violence, or alcoholism), depict tribes as though they are frozen in time, and often lack key context regarding culture and history, especially when conflict is involved (Loew, 2011; Larson, 2006; Weston, 1996). Other harmful stereotypes endure, including characterizations of American Indians as untrustworthy and animalistic (Tallent, 2013). Larson (2006, p. 109) provides a useful summary of specific stereotypes:

> Anti-Indian sentiment has a long history in newspapers that once explicitly promoted negative stereotypes. The terms "buck," "chief," "brave," "squaw," and "papoose" were common in newspapers until the 1960s. Stereotypes of the "noble savage," the "exotic relic," and the "degraded Indian" endured for decades in the news, adjusted to fit the events of the day. Reoccurring images "of Indians as exotic, warlike, childlike, or improvident" are still common.[1]

Broadly, the news tends to place Indigenous people into two categories:

> The "good Indian"/"bad Indian"...dichotomy found in films has also dominated news coverage of Native Americans. Who was "good" and who was "bad" depends on how much trouble particular Indians were giving whites in power at particular times. Neither stereotype considers Indians as complex and unique individuals.
>
> *(Larson, 2006, p. 109)*

Facile and misleading portrayals like these tend to linger indefinitely in the press, meaning that even today one can still find "culturally insensitive" news frames depicting Indigenous people as barbaric and lawbreaking, prompting Sanchez (2012, p. 9) to observe that "The mass media in the twenty-first century seem to remain focused on American Indian imagery that is struck in the eighteenth century."[2]

Importantly, entire frames often can be evoked by a single word, as recognized by Selene Phillips (2012): "After centuries in which the word 'Indian' has been part of our written and spoken languages, it is almost impossible to encounter the word without envisioning specific mental images" (p. 34). Along these lines, when writing about American Indians, journalists are forced to make "lexical decisions" that have significant meaning in terms of the way in which certain Indigenous groups are represented. In her study of the *New York Times* and *Los Angeles Times*, Seymour (2012) paid particular attention to four of the most frequently used words or phrases to refer to Indigenous groups in the U.S.: 1) broad and commonly used terms like "Native American" and "American Indian,"[3]; 2) the more pejorative, "Othering" term "the Indians"; the common "one-size-fits-all" referent of "tribes"; and the least commonly used "sovereign" or "nation" (p. 74). Why were these latter terms the least used? Seymour suggests that it may be because they connote a deeper sense of power and independence held by various tribal nations. Valerie Alia, who has focused much of her scholarship on environmental and social justice issues in relation to the media, makes clear in *Un/Covering the North* (1999) that there is continued "exoticization" of Indigenous groups in Canada by the news media. She also notes that what she terms Aboriginal-produced news is "marginalized at least twice – as *northern* news, and as *Aboriginal* news. In other words, it is marginalized first by region and then by culture" (p. 141).

One of the best, if older, summaries of the mediated depictions of Indigenous people comes from Weston (1996), who notes that these groups:

> have been patronized, romanticized, stereotyped, and ignored by most of mainstream America. The twentieth-century press has been complicit in this, seldom by design but certainly through the exercise of its conventions and values.... The very conventions and practices of journalism have worked to reinforce... popular imagery.
>
> *(p. 164)*

Alia (1999) recognizes that "the early history of communications about the North is steeped in the language of conquest and colonization. Despite centuries of change and decades of progress, that language persists in many of today's communications" (p. 13). For Alia, to study these communications is to understand the power of mediated representations, and she urges all journalists, regardless of background, to take responsibility for the "power and impact of their work.... I am extremely frustrated with the inadequacies of conventional training and practice, which remain grounded in ethnocentric principles of 'expert' authority and balance" (1999, p. 47). Things, of course, can and do change: when researching a conflict over a school's use of an "Indian" mascot, researcher Klyde-Silverstein (2012) discovered some improvements on the way in which Native Americans were covered by the press, noting that the press gave validity to their concerns over the mascot and did not portray them as "one-dimensional" beings. Tellingly, however, he also notes that although they became a powerful force in making news, "American Indians were not allowed to frame the issue themselves" (p. 119).

Framing scholarship on environmental justice

If the mainstream media outlets have tended to ignore or stereotype members of Indigenous groups, what happens when these groups make claims for environmental justice? While the previous chapter identified broader issues in the way the news media cover environmental justice issues, what does previous framing scholarship tell us about the specific frames associated with news stories involving Indigenous environmental justice issues? In short, Vickery and Hunter (2016) recognize a paucity of scholarly consideration regarding Indigenous environmental concerns. While some scholars (see especially Walker, 2011; Faber, 2008; Pellow & Brulle, 2005) provide broad discussions of how Native Americans' environmental justice struggles have been framed by the media, these works are fairly isolated and there are few in-depth studies. It is not that these concerns are uncommon or unimportant, but simply that they have not become a major focus of academic research to date. What this means is that Indigenous groups with environmental concerns have largely been ignored twice: by the news media first, and then by academia.

In a framing analysis of environmental justice cases involving Indigenous groups in my home state of Washington, I and one of my students examined media coverage of two Tribes: the Quileute and the Lower Elwha Klallam (Moore & Lanthorn, 2017). Based on an 1855 treaty, the Quileute Tribe was relocated to a small plot of land that was located in a tsunami zone. For decades (if not centuries), the Quileute desired to move its schools, playgrounds, and living quarters off the coastal plot to a safer spot inland. The Lower Elwha Klallam Tribe had endured the construction of a dam on their lands in the early 1900s that flooded ceremonial areas and significantly altered their fishing practice and livelihoods. What the two nations had in common was that the news media had ignored them for years: the Elwha were not mentioned during early coverage during dam construction, and the Quileute only began receiving scant coverage in 2006. Our findings revealed that while the local newspapers did not mention the Lower Elwha Klallam at all during dam construction in the early 1900s (a frame we termed *Omission*), by the time the dam was being dismantled in 2011, the Tribe had become recognized by the news media (which employed the twin frames of *Environmental Balance* and *Cultural Restoration*) as a key player in the process. While early frames for the Quileute included *Omission* and *Minimization*, later news coverage moved to a *Public Safety* frame, supporting the Tribe's move outside the tsunami zone. Overall, the research found that *Omission* (of Tribal concerns and voices) is a frame that constantly needs to be challenged.

Other scholarly work outside the U.S., including that by Jill Hopke (2012) on how the mainstream and "alternative" press in El Salvador framed the issue of gold mining, is particularly instructive, revealing that the mainstream press employed the rhetoric of *Corporate Rights* and *Benefits and Progress* to frame the discussion around the environmental impact of gold mining on Indigenous people – frames that privileged the corporate perspective over those of the communities most impacted by mining. The alternative press, meanwhile, framed the gold mining through the perspective of anti-mining activists and especially in terms of a *Community Rights* frame. The perspective of the alternative press outlets was thus that the government was negligent in caring for its citizens, thus highlighting a key difference in perspective based on news sources (Hopke, 2012).

In their analysis of environmental justice coverage in Canada, Deacon, Baxter, and Buzzelli (2015) found that environmental justice issues are often "muted" by Canadian daily newspapers, which "overtly ignore and subtly downplay through discourse environmental threats to marginalized local communities faced with pollution" (p. 419). The use of government sources in particular tended to frame environmental risk to communities of color as "unproblematic – rather than unjust" (p. 426). In addition, Deacon et al. found that the dailies often privileged the idea of economic growth over environmental justice concerns, emphasizing the risks of not protecting and nurturing financial prospects. Their findings instantiate what Haluza-Delay (2007) has observed, namely that "the environmental justice frame has not caught on in Canada" (p. 559).

What the previous research on environmental justice coverage by the news media around the world makes clear is that first, there is a relative paucity of press attention paid to Indigenous environmental concerns and issues; and second, when attention is paid to environmental justice issues, Indigenous voices often are stereotyped, marginalized, and/or minimized. The question for this chapter then becomes how the Standing Rock Sioux Tribe's struggle was framed by the news media. Did the news media continue frames of "omission"? Did they negate or minimize the Tribe's concerns about the need for clean water? What has changed, and what remains the same when it comes to the image in the press of Indigenous groups like the Standing Rock Sioux Tribe? Before answering these questions and presenting the results of this research, the methods of analysis are described in the next section.

Interpretive framework: framing as method and theory

Outlined in Chapter Two, this research is guided by several normative theories regarding the press. First, there is the idea regarding *Social Responsibility* that arose out of the Hutchins Commission, which laid the foundation for definition of a public-oriented, responsible press. In this view, the news media should demonstrate a strong and unwavering public-service function and avoid undue corporate influence. In addition, several scholars have put forth a democratic theory regarding the media, which contends that the press should: be a watchdog for the public, holding those in power (including corporations and government) accountable; work to discern truth from lies; "regard the information needs of all people as legitimate; if there is a bias in the amount and tenor of coverage, it should be towards those with the least amount of economic and political power"; and present a balanced range of perspectives and opinions on important issues (McChesney, 2014, p. 13).[4,5] Thus, the results of this analysis are meant to lead to an understanding of whether the coverage of the struggles around the Dakota Access Pipeline adhered to these normative principles through understanding what frames were used. As such, the normative theories are intended to complement the theoretical implications inherent in framing itself.

Frames are, to put it bluntly, incredibly common. They are used not simply by the professional news industry but also by special interest groups, politicians and political groups, numerous types of organizations, and individuals (Lakoff, 2010), and as such are employed in what is considered to be "everyday" communication as well as in "academic discourse"[6] (Snow & Benford, 1992, p. 136).

Working specifically in the academic arena, Gamson and Modigliani (1989) provide a broad, working definition of framing:

> media discourse can be conceived of as a set of interpretive packages that give meaning to an issue. A package has an internal structure. At its core is a central organizing idea, or frame, for making sense of relevant events, suggesting what is at issue.
>
> *(p. 3)*

Accepting that frames serve as a "central organizing idea" paves the way for a deeper understanding of their function – namely, that they "provide meaning to an unfolding strip of events, weaving a connection among them. The frame suggests what the controversy is about" (Gamson & Modigliani, 1987, p. 143). Snow and Benford (1992) expand upon this notion to define frames as "interpretive schemata that simplifies and condenses the 'world out there' by selectively punctuating and encoding objects, situations, events, experiences, and sequences of actions within one's present or past environment" (p. 137). Seen from these broadly similar perspectives, frames can be seen to establish and organize meaning (Gitlin, 1980), they tend to simplify complex issues and events (Snow & Benford, 1992), and because of their near ubiquity, often can be evoked with a single word or phrase (Lakoff, 2010). Because of this, framing in this research is seen to provide both method and theoretical framework that complements normative theories of the press.

Framing analysis of Standing Rock: what criteria are included?

To make explicit how framing is understood and used as both method and theory in this research specifically, it is useful to begin with reference to an oft-cited quote from Entman, who contends that frames function to "*define problems...diagnose causes...make moral judgments...and suggest remedies*" (p. 52, emphasis in original). In other words, frames do important work in four parts. First, they are integral in defining the scope of an issue through definition of the problem, outlining the dimensions, and identifying who is involved and/or impacted. Second, frames provide indications as to what (and who) has caused the problem at hand. Third, frames often furnish moral interpretations and evaluations of the problem, especially regarding who or what may be harmed by the issue or event. Finally, frames contain potential solutions, including who is responsible for mitigation of a problem or an issue.[7,8] The framing analysis conducted for this study, then, pays particular attention to how the news media cover these four dimensions.

Within this broader interpretive framework are more specific points of assessment that are utilized here. Gamson and Modigliani (1989) provide a systematic way to evaluate the framing of a text by identifying five distinct *framing devices*:

visual images; prominent examples and points of historical comparison; metaphors; catchphrases (specific phrases or slogans); and depictions (pp. 3–4). In addition, although Ibarra and Kitsuse (1993) do not identify their work as framing (but instead as constructionist in the study of social problems), their criteria for evaluating a text also proves quite useful for framing analysis in that they consider:

1. *rhetorical idioms*: the use of what they term "moral vocabularies" that make a claim for significance while exhibiting "hierarchies of value" (pp. 31–32);
2. *counterrhetorical strategies*: including "sympathetic" and "unsympathetic,"[9] these are approaches to an issue or event that "block either the attempted characterization of the condition-category or the call to action, or both" (p. 38) in a "yes...but" format (p. 41);
3. *motifs*: recurrent themes or "figures of speech" within a text (p. 43);
4. *claimsmaking styles*: including socially recognized deliveries of claims, like "scientific style," comic, theatrical, civic (pp. 47–48);
5. *setting*: the setting within which the communication takes place (they note that "*the media* constitute one class of setting" and include legal-political settings like courts as well) (p. 50).

Adapting the criteria laid out by previous research, the framing analysis for this study will place special emphasis on: 1) photos and images used with the stories; 2) the specific language used in the articles, including particular phrases and slogans as well as adjectives used to describe the actions and characters of protesters, police, the local public, and government; 3) prominent and recurring examples used in the stories; 4) the tone and style of the articles; and 5) the sources quoted most commonly. It is difficult to overstate the importance of language in framing, because even seemingly value-neutral words can activate particular frames (Seymour, 2012; Lakoff, 2010). One important example of language occurred during the conflict over the Dakota Access Pipeline, when law enforcement increasingly used the term "riot" to describe the actions of protesters. How news agencies responded to this labeling is important – did they adopt or reject this value-laden language?

Identifying frames thus enables an understanding of how a journalist or news agency is choosing to present the Indigenous struggle over the pipeline – from causation and origins to responsibilities and solutions. Boykoff (2006) notes that "by framing socio-political issues and controversies in specific ways, news organizations present – if tacitly – the foundational causes and potential consequences of a social problem or issue, as well as possible remedies" (p. 204). By this same token, frames can essentially determine whether or not an issue even becomes recognized as a legitimate "problem" that requires solving. Defining framing in this way touches directly upon the constructionist perspective when it comes to social, political, or environmental issues, because

> problems do not become recognised or defined by society as problems by some simple objective existence, but only when someone makes claims in

public about them. The construction of a problem as a 'social problem' is then largely a rhetorical or discursive achievement, the enactment of which is perpetuated by claims-makers, takes place in...public arenas.

(Hansen, 2010, p. 28)

The constructionist approach to framing is particularly useful here, for one can query whether the media considered #NODAPL to be a problem worth identifying and exploring. The fact that the mainstream media had ignored what was happening at Standing Rock in 2016 and 2017 invites questions about what and who the news media industry defines as being newsworthy. And while the analysis in this chapter focuses on the pivotal moment that mainstream media coverage did begin in earnest (after the viral video by *Democracy Now!* on Labor Day in 2016), the absence of coverage prior to that date (and even, for many media outlets, after this time) must be considered as part of the analysis. As Entman notes, "frames are defined by what they omit as well as include" (1993, p. 54) making omission one of the most powerful frames.

Engaging with the Standing Rock Sioux Tribe's Environmental Justice framing of #NODAPL

It is important to note that while the majority of framing research (including this one) places specific focus on the news media, George Lakoff (2010) reminds us that frames are adopted and utilized by a variety of people, organizations, and movements: "successful social movements require the coherence provided by coherent framing. Think of the union movement, the anti-war (or peace) movement, the civil rights movement, or the feminist movement. The basic ideas are simple and straightforward" (p. 80). What Lakoff recognizes is twofold: that frames are incredibly common and used by many different organizations and entities; and that, in order to gain much-needed traction and public support, any movement (be it environmental, social, or political) needs to employ clear, expressive frames.

Since the #NODAPL movement became widely known in the fall of 2016, the early scholarly studies that have emerged recognize that the Standing Rock Sioux Tribe provided their own powerful environmental justice framing for their struggle. Falc (2017) in particular has interpreted the phrase "Mni Wiconi" ("Water Is Life") as being deeply evocative, which she sees as enabling the movement to gain quick and widespread support. Similarly, in their analysis of the Comanche Trail Pipeline protests in Texas, Alanis-Ramirez and Tena (2017) propose that future environmental justice movements should follow in the footsteps of Standing Rock by creating effective frames that provide both an effective solidification of message and a powerful rallying cry for their cause. Both studies found that the Indigenous message of "Water Is Life" was difficult for the opposition to contradict, for who would argue against the importance of clean water for sustaining life? Ibarra and Kitsuse (1993) likely would perceive a "hierarchy of value" in this catchphrase, "with which it is difficult to disagree without discrediting onseself" (p. 38). While the analysis that follows here (like many scholarly assessments of frames), focuses on the specific schema used by the news media, it also recognizes the power of the frames created by the "water protectors" themselves. Thus, a key part of the analysis becomes how (or, indeed, whether) the news media organizations in question engaged with the #NODAPL rhetoric and the language of protest employed by the Tribe.

Selection of newspapers and countries

This chapter, for the sake of organization and brevity, presents the results of newspapers within the U.S., while Chapter Five provides the findings for newspapers in Canada. In the U.S., analysis focused on the *New York Times*, the *Bismarck Tribune*, and the *Indian Country Media Network*; in Canada, the research included the *National Post*, the *Globe and Mail*, and the *Calgary Herald*. The choice of these papers was guided by several factors, but primarily the ability to compare results, especially by country and by region. Newspapers from the U.S. and Canada were chosen so as to be able to discern similarities and differences in how newspapers in each country – both of which are involved in fracking and who have large Indigenous populations – would frame the issue of the Standing Rock struggle against the oil and gas industry. Of particular interest was gaining the ability to understand how local coverage (e.g., the *Bismarck Tribune* and *Calgary Herald*) differed from national coverage (the *New York Times* and the *National Post*), as well as assessing how mainstream, commercial journalism would compare to Indigenous-produced media like the *Indian Country Media Network*. Importantly, selection was not based on quantity: while the U.S. the *Bismarck Tribune* had 425 news stories, the *New York Times* had 45 articles, and *Indian Country Media Network* had 46, newspapers in Canada published many fewer articles. The less frequent coverage by Canadian newspapers were included in the analysis for two reasons: first, because the frame patterns were still clearly discernible, and second because omission itself is considered a significant element in framing analysis. The framing analysis covers variable time periods for each newspaper that depended upon when the daily began their coverage, and ends in March 2017 when the majority of protests had died down. The *Bismarck Tribune*, however, continued its coverage past the time of this writing because it was covering court cases and the operation of the pipeline.

Articles for almost all newspapers were identified through a search of the words "Dakota Access Pipeline" in *Lexis Nexus* (now *Nexis Uni*). The results from that search were cross-referenced with the *Access World News* (*AWN*) database when possible in an effort to locate all articles. While these two resources provided full-length articles, they did not contain images, and so one aspect of the analysis involved accessing individual articles at the source (e.g., the websites of each newspaper) to see the images they used in conjunction with each story. Selection of articles for *Indian Country Media Network* (*ICMN*) took a slightly different path, in that neither *AWN* nor *Nexis Uni* included this news source in their database. As a result, I found articles directly from the website of *ICMN*. And while the search results using "Dakota Access Pipeline" said that they yielded 1,303 articles, the search results appeared to stop at page ten. As a result, all 43 articles were chosen from available pages. In addition, as of this writing (in early 2018), *ICMN* has the following "Editor's Note":

> As of September 4, 2017, Indian Country Today Media Network, publisher of Indian Country magazine and IndianCountryMediaNetwork.com, is taking a hiatus to consider alternative business models. The staff of ICTMN has been honored to serve the best audience they could have possibly imagined, and direct all attention to the following letter from publisher Ray Halbritter. During the hiatus, new posts, new magazines and new books will not appear on the site and email newsletters will not be sent while we consider a new way forward. The site will remain accessible and maintained in its current form through January 31, 2018.

Analysis included only articles written directly by each newspaper: for example, if the *Bismarck Tribune* picked up a story from *Reuters*, those articles were excluded. This was consistent (when and where discernible) for all news sources. Opinion pieces were not considered (as they are almost always written by non-journalists), but editorials expressing the views of the newspapers editorial staff (which were small in number) were included, especially because, in the words of the *New York Times*, they most clearly reflect "the voice of the board, its editor and the publisher."[10] In addition, articles typically had to be longer than 100 words (in some short articles it was difficult to discern a frame), but if a frame was identifiable then the article was included.

The results of the framing analysis for newspapers within the U.S. are presented in the pages that follow.[11] Definitions are provided for all frames: if the frames are already established in existing scholarship, original definitions are provided; if the frames were relatively unique to this research (having emerged from the current analysis), I have provided the parameters of the new frame with examples.

In addition, some notes are in order. First, although article frames are tallied, this research should not be considered quantitative, for no statistical tests were performed and no numbers truly "crunched": counting the number of frames per each paper simply was an aid to understanding which frames were the most frequently employed by each paper. As such, the research remains far less focused on numbers and more on overall trends and recurrent frames per each paper and each country. This position is underscored by the relatively subjective method of framing: while I believe that other scholars would likely discover the same broad trends in framing for each newspaper, when it comes to the specific count and naming of frames, it is highly likely that another scholar would have somewhat different findings.

Finally, a note about the organization of the framing results is in order. Although initially I had planned to present the frames for each paper in one grouping, the flow of the frames during different stages of #NODAPL made it more intuitive to break the papers' coverage into time periods. For the *New York Times*, there were two natural "breaks" where framing distinctions could be made. For the *Bismarck Tribune* (with 425 articles), there were four distinct frame grouping per time periods that reflected different positions by the paper. As such, what you will find is what I feel is the best way to present a complex set of frames throughout a decently long time span of pipeline conception, construction, and completion – with various stages of resistance and activism along the way.

Analysis of U.S. coverage – *New York Times*

The *New York Times* (*NYT*) is a national, seven-day newspaper that, per its 2016 Annual Report 2016,[12] was circulated in 195 countries and had roughly 2.9 million paid subscriptions in both digital and print format. Within the U.S., the report notes that it

> had the largest daily and Sunday circulation of all seven-day newspapers for the three-month period ended September 30, 2016, according to data collected by the Alliance for Audited Media ("AAM"), an independent agency that audits circulation of most U.S. newspapers and magazines.

In addition, a *Pew Research Study* noted that in 2016, the paper added "more than 500,000 digital subscriptions in 2016 – a 47% year-over-year rise" (Barthel, 2017).

Analysis of the paper's coverage of the pipeline included 45 articles spanning from 2011, when the general idea of the pipeline first emerged to transport Bakken oil out of North Dakota. Consistent coverage continued until March 2017, after which time the paper published stories more sporadically. Many articles that contained the keywords "Dakota Access Pipeline" after March contained stories relating to major court cases or mentioned the pipeline project tangentially, especially in relation to the other major pipeline (Keystone XL) being considered at this time. After late February, stories were sporadic and mostly came in the form of op-ed pieces.

The results for this newspaper are grouped into two sets of articles: those written prior to Labor Day in early September, 2016, and those written after Labor Day, when *Democracy Now!* posted its viral video of Indigenous protesters being bitten by dogs and pepper sprayed. In the five years leading up to the construction of the Dakota Access Pipeline (2011–2016), there were a a relatively small number (eight) of articles that covered the pipeline; after Labor Day 2016 up until early March 2017 (about a six-month period), there were 37 articles. Noted earlier, not only did the frequency of coverage increase, but, unexpectedly, the frames shifted as well.

New York Times – *before Labor Day 2016*

Although small in number, the frames varied widely (perhaps due to the wide time frame in which they were written). In the articles prior to Labor Day, two framed the North Dakota pipeline as a clear boon to U.S. energy independence (the *Energy Independence* frame), two defined the pipeline through the lens of profit to be made (*Economic*), two described the potential pitfalls of the pipeline to individual citizens (*David and Goliath*), one defined the protesters as dangerous and confrontational (*Law and Order*), and one took a "balanced" tone that attempted to present all sides of the burgeoning conflict (*Professional Journalism*). With the exception of *Energy Independence*, all of these frames recur throughout this paper and others in the U.S. and Canada.

Energy Independence

Energy Independence perspective, originally defined by Gamson and Modigliani (1989) as a way to frame a "pronuclear argument," is meant to underscore the importance of American self-sufficiency. Adapted to this analysis, this frame makes a clear argument for pipelines as a necessity for the U.S. to become independent by using its own oil reserves. Articles that used this perspective even used this exact language. In an early article on November 13, 2012 (titled "Making an Energy Boom Work for the U.S.") the journalist noted that the rise of domestic fossil fuel production would have significant, far-reaching changes:

> Experts, for instance, argue that an increase in North American energy self-sufficiency will lead to equally drastic changes in geopolitics. Suddenly having a great wealth of domestically produced gas and, increasingly, oil, the argument follows, will allow the United States to look inward and take less interest in international affairs....

The article suggests how pipelines from North Dakota will play into energy independence, including increased security of transport. The image associated with the article was a set of playing cards placed over a map of the world – the cards seem to suggest a global game when it comes to fossil fuels, with some cards face down (seemingly to indicate uncertainty). On October 24, 2012, another article ("Bigger Than Either of Them?") made a similar claim, putting a positive spin on the high domestic production of oil and gas, noting that "with domestic oil production climbing rapidly since 2008, oil imports have been cut to just over 40 percent of domestic supplies, the lowest level in two decades." The language used – including "success," "self-sufficiency," and "secure" serves to frame the pipeline as a positive development.

Economic

The other common frame during this time period was *Economic*. This frame defines the pipeline through the lens of money – from speculating about the profits to be made as well as potential losses suffered if pipelines are not built. This frame thus defines the North Dakota pipeline as integral to economic health and national economic growth. An article on December 16, 2011 ("Kinder's Major Bet on a Boom"), strongly equates prospecting in oil pipelines with the language of striking it rich by highlighting Richard Kinder, CEO of Kinder Morgan, who has "personally made billions of dollars operating the industry's equivalent of a toll road: pipelines. Hydraulic fracturing techniques – despite causing a growing controversy – are creating a once-in-a-generation boom in oil and gas drilling in the United States." Here, the article uses the example of an affluent pipeline executive made even wealthier by investing in fossil fuel infrastructure, while suggesting that others can do the same due to ample "opportunities" in the business. An October 25, 2012 article titled "Oil Refining's Fortunes Rise" uses the language of economic benefit, noting that investment in "new pipeline terminals" is paired with profits "flowing at most refineries."

Although small in number, *Energy Independence* and *Economic* frames seemed to go hand in hand, with articles categorized under one frame still exhibiting the language and characteristics of the other to a lesser degree. The article "Making an Energy Boom work for the U.S." exemplifies this pairing, noting that "the rapid rise in output of natural gas and, more recently, oil in the United States is transforming the country's energy and economic landscape. These production booms are revolutionary." Here, the article goes on to note that the "revolution" taking place is energy independence as well as profit.

Devil's Bargain

Although many of the paper's early articles appeared to breathlessly tout fossil fuels – and the pipeline infrastructure needed for them – as a desirable opportunity, this view was tempered somewhat by two articles that appeared in 2014 – the last of the articles on the Dakota pipeline until 2016. On November 23, while an article titled "Downside of the Boom" still employed language evocative of a profitable "gold rush" it also provided an environmental perspective regarding the danger of oil spills by quoting a farmer who exclaimed that "the gold is flowing – Bakken gold." By "gold is flowing," he meant that oil was spilling. The article places responsibility on the North Dakota government for the environmental costs associated with oil and gas production, noting that

> state leaders rarely mention the underside of the boom and do not release even summary statistics about environmental incidents and enforcement measures.... "We're spoiling the child by sparing the rod," said Daryl Peterson, a farmer who has filed a complaint seeking to compel the state to punish oil companies for spills that contaminated his land.

The images placed throughout the article echoed the disapproving tone, including an oil worker who was badly burned during an accident, and a farmer taste testing water on his land after it was exposed to wastewater from the accident. The caption notes that the contaminated water left his land "useless."

An article published the day after, titled "Where Oil and Politics Mix," placed focus on a different downside to the North Dakota oil "boom": landowners' rights. The article notes that because "mineral rights trump surface rights, this made many residents of western North Dakota feel trampled once the boom began." In other words, the story suggests that the oil beneath the surface was more important than the rights of individual landowners to use the surface of their land as they saw fit. The story was peppered with images of the harm associated with oil transportation, including a derailed and burned oil train that was juxtaposed with an older landowner walking through his fields wearing denim and a baseball cap, accompanied by his labrador retriever.

Both of these articles – highlighting the various dangers associated with fossil fuel infrastructure in North Dakota – demonstrate a *Devil's Bargain* perspective. Originally defined by Gamson and Modigliani (1989, p. 25) as a "thoroughly ambivalent package" that is simultaneously "for and against" nuclear power, it is adapted here to recognize that the newspaper seemed to have contradictory messages in these types of articles: on one hand, both articles use the language of an economic "boom" (one equating oil with gold), which could be seen as a positive take on fracking and pipelines. On the other hand, the articles underscore the significant dangers – to livelihoods, to health and safety, and to farmers' rights – that accompany the industry. Thus, this frame recognizes the real detriments that accompany the benefits.

Law and Order

After late 2014, there were no articles on the pipeline until late August, 2016. One of the articles ("Tension on the Plains as Tribes Move to Block a Pipeline") was different from others because it had none of the earlier frames, but instead a new one that likely was due to new developments in the pipeline – namely, the growing resistance from the Standing Rock Sioux Tribe and others who joined them. The article describes how the protest began that day: "horseback riders, their faces streaked in yellow and black paint, led the procession out of their tepee-dotted camp." Accompanied by the "Othering" language (where the protesters were portrayed as anachronistic, primitive, and warlike), the article exhibited a *Law and Order* frame. This frame typically depicts protesters as dangerous and often irrational, primarily uses the quotes and perspective of law enforcement, and often focuses on the number of protesters arrested (Moore & Tucker, 2017). A section from the article demonstrates this frame:

> More than 20 people have been arrested on charges including disorderly conduct and trespassing onto the construction site. The pipeline company says it was forced to shut down construction this month after protesters threatened its workers and threw bottles and rocks at contractors' vehicles. Sheriff Kyle Kirchmeier of Morton County…said at a news conference that he had received reports of weapons and gunshots around the demonstration, and that protesters were getting ready to throw pipe bombs at a line of officers.

This article suggests that the Indigenous protesters were violent and unpredictable and, in concert with the language of primitivism, evokes the facile "cowboys and Indians" trope. One quote in particular highlighted this: "This month, a line of sheriff's officers retreated in the face of riders on horseback circling and yipping through the grass." While the article does give space for the perspective of the Standing Rock Sioux Tribe (it was noted that the action could be "a Lakota gesture of introduction"), the reader still is left with an uneasy image of violent "primitives" on the prairie. While one doesn't want to make too much of a single article, the story's significance is underscored by the fact that this was the first description by the paper of the growing resistance.

Professional Journalism

The last article prior to Labor Day – "North Dakota Oil Pipeline: Who's Fighting and Why?") – was published on August 26. Written in a balanced and objective tone that presented all sides (the hallmark of professional journalism), it includes multiple perspectives, including the pipeline and the protesters. The story also included multiple images, including an Oglala Lakota girl and her mother holding signs and pipes being stacked in preparation for installation, the article had headers like "What does each side want?" and "What is the latest from North Dakota"? The article did end on an interesting note, however, observing that "more than half of the country's pipelines [a]re at least 50 years old," an age that "put[s] public health and the environment at risk."

Summary of pre-Labor Day coverage

The placement of the articles in each paper yielded some patterns, in that all of the articles with the *Economic* and *Energy Independence* framing were placed (perhaps unsurprisingly) in the "Business" or "Energy" sections. On the other hand, the articles exhibiting *Devil's Bargain* made the front page, seeming to underscore the importance the paper placed on the rights of certain individuals (including farmers, landowners, and citizens) as well as the importance of protecting the environment against oil and gas developments and infrastructure. The one article demonstrating *Law and Order* was placed on the back pages of the first section of the paper.

Noted earlier in this chapter, considering what information is omitted is essential during framing analysis. Here, one first could consider the relatively small number of articles on the pipeline itself, even as protest activity began to build in early 2016 when the Tribe found out that the pipeline would be built upstream from their primary water source. In particular, many Indigenous youth participated in a multi-state run for hundreds of miles as part of "ReZpect Our Waters" in early April of 2016 to draw attention to the Standing Rock plight. The paper did not cover this. The fact that the newspaper included only two articles prior to Labor Day about the impact of the pipeline on the Tribe's water supply suggests a crucial absence in regard to Indigenous perspectives and voices prior to conflict-driven events of

September, 2016. As a result, while clear frames were identifiable from 2011 to 2016, the most important frame during this period is *Omission*, as the concerns of the Standing Rock Sioux Tribe were conspicuous in their absence.

New York Times – *after Labor Day 2016*

In the six-month period from early September, 2016 to February, 2017 the *NYT* published 37 articles, roughly five times as many articles as it did in the five-year period prior to this. The primary frames discernible in this period of time, in order of frequency, were *Environmental Justice* (20), *Conflict* (11), and *Professional Journalism* (6). What emerges is in this time period, then, is not simply a greater number of articles, but a very different set of frames.

Environmental Justice

The most common frame seen during this time frame was *Environmental Justice*, with the first example seen on September 9 – a few days after the highly publicized viral video from *Democracy Now!*. The definition of this frame used for this research pulls from Walker's (2011) description of an environmental justice frame that has emerged within the U.S., which includes a consideration of: 1) "the politics of race" – stemming from the beliefs and actions emerging from historic civil rights activism; 2) the idea of "justice to people" in relation to their environmental concerns; 3) an expanded view of environmental harm, making room for more concerns and risks than earlier defined; 4) *participatory justice* (who can be involved) as well as *distributive justice* (e.g., "who gets what"); and 5) the responsibility of corporations and government, including an emphasis on institutionalized forms of inequality and discrimination[13] (pp. 20–23). Defined in this research, then, the *Environmental Justice* frame recognizes race as a factor in the #NODAPL struggle, emphasized the need for justice for the Standing Rock Sioux Tribe regarding both clean water and a place at the decision-making table, and paid close attention to the role that Energy Transfer Partners and the state and national government played in environmental justice issues.

In the September 9 article, titled "I Want to Win Someday: Tribes Make Stand against Pipeline," the treaty rights and historical injustices are highlighted in their importance:

> History, like a river, runs deep here. And residents… say the pipeline battle has dredged up old memories and feelings about lost lands and broken treaties with the United States government, as well as their worries about the future of land and water they hold sacred.

While the article begins with the "cultural catastrophe" created by the Oahe Dam project that inundated Sioux lands, it ends by tying that past injustice into a future one: the Dakota Access Pipeline. The article continues by quoting Dave Archambault II, then-Standing Rock Sioux chairman as saying, "The trauma we deal with today is a residual effect of 1958, when the floods came." The image associated with the article portrays Verna Bailey, whose grandparents' home was lost due to the dam, walking through an open field.

Articles that contained this frame focused broadly on the historical treatment of Native Americans as context for understanding the Tribe's concerns about access to clean water. In "Water, Water Nowhere" on October 20, the article notes that

> Most Americans take safe water for granted: Turn the tap, and there it is. But recent protests against the Dakota Access pipeline on the Standing Rock Reservation in South Dakota are a reminder that some Americans still worry every day about having enough clean water to survive.

As this passage makes clear, clean water is an environmental justice issue that disproportionately impacts people of color and is a human right.

The *Environmental Justice* frame seems to take firm hold in early November with two pieces from the editorial board. The titles of the editorials indicate the content within well enough, but some of the passages are well worth reproducing here. On November 3, in "Time to Move the Standing Rock Pipeline," the board wrote that the protests represented

> an intensely bitter confrontation that came wrapped in historic injustice and seemed destined to end in grief. The $3.7 billion Dakota Access pipeline... would not enter tribal land but it would pass close enough for the Sioux to fear grave damage from a leak or spill. Its current proposed route runs less than half a mile north of the reservation and under the Missouri River, a source of drinking water. Though the pipeline would mostly cross private property, the tribe also argues that these have been the Sioux's ancestral lands since antiquity, and construction would damage sites of deep cultural and historic significance, including burial grounds.

The editorial board then seemed to make reference to its earlier stance – namely, that pipelines are a relatively safe way of transporting fossil fuels, but now contends that the Dakota Access Pipeline can't be justified:

> A pipeline may well be the most profitable and efficient way to move a half-million barrels of crude oil a day across the Plains. But in a time of oil gluts and plummeting oil prices, is it worth it? Is it worth the degradation of the environment, the danger to the water, the insult to the heritage of the Sioux?

While the board's comment about low oil prices invites a question (namely, would they deem the damage to the environment and the Tribe "worth it" if oil prices were higher?), it is clear that that the editorial's question is rhetorical: that is, their stance suggests that having more oil is not worth the environmental injustice against Indigenous nations. The image accompanying the editorial was a sketch of an oil pipeline in profile, dripping black oil onto a countryside with trees and rolling hills.

On November 23, in "Power Imbalance at the Pipeline Protest," the paper's editorial board again demonstrates the *Environmental Justice* frame the most clearly and forcefully:

> When injustice aligns with cruelty, and heavy weaponry is involved, the results can be shameful and bloody. Witness what happened... when law enforcement officers escalated their tactics against unarmed American Indians and allies who have waged months of protests against the Dakota Access oil pipeline. They drenched protesters with water cannons on a frigid night, with temperatures in the 20s. According to protesters and news accounts, the officers also fired rubber bullets, pepper spray, concussion grenades and tear gas. More than 160 people were reportedly injured, with one protester's arm damaged so badly she might lose it. The department's video was meant to portray the protesters as dangerous troublemakers, but the photos and videos in news reports *suggest a more familiar story – an imbalance of power, where law enforcement fiercely defends property rights against protesters' claims of environmental protection and the rights of indigenous people.* American Indians have seen this sort of drama unfold for centuries – native demands meeting brute force against a backdrop of folly – in this case, the pursuit of fossil fuels at a time of sagging oil demand and global climatic peril. [*emphasis added*]

By focusing on the tactics employed by the Morton County Sheriff's Office against the protesters, the paper's editorial board clearly condemns what it feels is unjustified force that is resulting in significant harm. The piece references the "department's video," which was released by local law enforcement to apparently show the "truth" of the protests, but, in the view of the paper, the truth that it revealed spoke more to police brutality against people of color than it did to the violent nature of those who participated in the struggle for access to clean water. The image at the top of the editorial was a Reuter's photo of tear gas spilling across a barbed wire fence crossing the countryside. People with guns and demonstrators are seen in dark silhouette while the caption below reads: "Police used tear gas on demonstrators opposed to the Dakota Access oil pipeline in North Dakota, on Sunday."

In regard to environmental justice concerns, then, the newspaper presents two injustices: the first is the endangerment of the Tribe's water supply, the second pertains to their treatment during protest. To use the phrasing of Walker (2011), the paper seemed to focus on both *distributive justice* (do the Standing Rock Sioux have access to clean water?) as well as *participatory justice* (who gets to sit at the table and be an integral part of the discussion about important environmental justice issues?).

Conflict

During this time period, the second most common frame was *Conflict*. The conflict frame has been studied extensively in political communication scholarship, with the definitions and discussions being useful in this context as well. Valkenburg and Semetko (2000, p. 95) provide a clear definition of this frame, which "emphasizes conflict between individuals, groups, or institutions as a means of capturing audience interest." Adapted to this research, articles that contained the *Conflict* frame highlight the pipeline protest as an acrimonious clash primarily between "water protectors" (a term used by the Standing Rock Sioux Tribe but not adopted by the paper, which most often referred to them as "protesters") and law enforcement. When presented on its own, the frame does not appear to privilege any one side. Bartholomé et al. (2015, p. 439), who write on political conflict specifically, identify the complexities that can be involved with this frame, observing that "Journalists decide if and how to report about political conflict. They may seek out political conflicts, amplify political conflicts for the attractiveness of the story, or even actively orchestrate and manufacture conflict frames." Because conflict became an unavoidable component of the *Environmental Justice* frame, it is important to distinguish them here, because not all articles focusing on tension and struggle exhibited a concern for justice but instead focused solely on the struggle itself. Articles categorized as *Conflict* were those whose sole focus was placed on the nature of the conflict itself. In addition, while the definition of *Conflict* used for this research does not necessarily assume journalistic magnification of the struggles between the protesters and the police, it does recognize the role that journalists play in how a conflict is constructed for news audience.

Examples of the *Conflict* frame are numerous. An article in late October, titled "Tension over Pipeline Hits Boiling Point," presents it this way:

> For months, tensions had mounted between protesters and law enforcement officials over the fate of an oil pipeline not far from the Standing Rock Sioux Reservation. Late this week, the strained relations boiled over as officers tried to force the protesters out of an area…. Scores of officers dressed in riot gear walked in a wide line, sweeping protesters out… as face-to-face yelling matches broke out. Several vehicles, including at least one truck, were set ablaze. A standoff unfolded beside a bridge known as the Backwater Bridge, where protesters set fire to wooden boards and signs and held off the line of officers over many hours.

The use of language like "boiling over," "riot gear," and "standoff" evokes the idea of violent fights with anticipated escalation. The video accompanying the article, which used the voices of those protesting (including members of the Navajo, Oglala Nation, as well as Indian Country Today journalists) showed armored vehicles with concrete barriers on a road and heavily armed law enforcement handcuffing protesters who were on the ground.

An article titled, "US Suspends Pipeline Work in Tribes' Path" on September 10 similarly paints a picture of high tension:

> In recent days, protesters have clashed with the pipeline company's contractors and private security guards, and officials in North Dakota have

stepped up patrols and warned of rising tensions as ranchers, sheriff's officers, tribal leaders and protesters waited for a ruling on the Standing Rock
Sioux Tribe's federal lawsuit to block construction on the pipeline.

This phrasing is interesting because it positions the protesters as creating conflict
with the oil company and law enforcement, and not the other way around, but
the article stops short of containing a *Law and Order* frame due to the numerous
quotes from spokespersons representing both the Standing Rock Sioux and the
Indigenous Environmental Network, helping to soften this frame through their
voices and perspectives.

Other articles noted that tensions were rising and that

ranchers are arming themselves before they climb onto tractors or see to
their livestock. Surveillance helicopters buzz low through the prairie skies.
Native Americans fighting to prevent an oil pipeline near the Standing
Rock Sioux Reservation are handing out thick blankets and coats.

(October 11)

A November 2 article contended that the "prairie is seething." The fact that the *NYT*
placed heavy focus on tensions and skirmishes between protesters and law enforcement is unsurprising, as conflict is often seen as a common element of news stories.

Professional Journalism

The third most common frame was *Professional Journalism*. This frame is defined
as creating a veneer of objectivity by the journalist in a "he said, she said" manner that provides little context and relies heavily on official sources (Schudson,
2011). Another way of conceiving of this frame is through the notion of "balance" (discussed in more detail in Chapters Two and Three). The best example
comes from an article in early November that was tellingly titled, "The View
from Two Sides of the Standing Rock Front Lines." One side was provided by
an Indigenous activist (Horinek) from the Ponca Tribe who asked rhetorically,
"Don't you drink water too?" The accompanying image showed him walking
against the backdrop of a blue sky. The article then progressed to portray the
other side – that of local sheriff (Moll), who was quoted as saying "folks are terrified." The image next to his perspective showed him next to his squad car with
the backdrop of the prairie. After providing his perspective, the article ends.

Articles like these from the paper are emblematic of the *Professional Journalism*
frame because they only provide readers with two clearly opposing sides. It
also reveals some of the difficulty in determining frames during this period of
coverage, because there were several articles that initially seemed to contain a
justice framing: that is, they provided the perspective of the Tribe and made
a case for considering their needs. But when that perspective is simply countered with another, opposing view, it constricts an argument for justice and

instead highlights the benefit of seeing all sides. That, of course, is the point of "balance," because the paper is attempting to look impartial. That stance changed, however, as the pipeline struggle grew more violent – the paper largely dropped its professional stance and embraced the environmental justice perspective.

Summary of New York Times framing analysis

Summarizing the framing analysis of the *New York Times'* coverage of #NODAPL, it is clear that the frames shifted significantly after the events that occurred on Labor Day weekend. Intriguingly, the *NYT* never made explicit mention of the viral *Democracy Now!* video showing Indigenous people being maced, hit, and attacked by dogs. Instead, the closest reference to the events of Labor Day was an October article noting that the "riot charges" leveled at reporter Amy Goodman by local law enforcement during her filming of the viral video were dropped by a judge.

Although the newspaper only obliquely referenced the events of Labor Day, it would appear that those events shaped, at least in part, the way in which it covered the pipeline protests, shifting ultimately to a frame emphasizing the need for justice. This pivot ultimately suggests that the paper was responsive to the Standing Rock Sioux, and that the Tribe may have been key in shaping the framing of the paper. This is especially evident in the fact that the paper published at least two opinion pieces from then-tribal chairman Dave Archambault II: pieces that emphasized the need to respect the Tribe's ancestral lands ("Taking a Stand at Standing Rock") and the broader environmental, cultural, and political consequences of the pipeline ("An Indian Protest for Everyone").

The paper's focus on the need for environmental justice, however, is complicated by two aspects of coverage. The first is the paper's focus on "balance." The myriad problems associated with the veneer of objectivity are discussed in more detail in Chapter Three, but a quote from Nick Davies' *Flat Earth News* book reveals the problem succinctly:

> Balance means never having to say you're sorry – because you haven't said anything…Applied to statements of fact, the rule is the embodiment of neutrality, and with the same result, that journalists are encouraged to abandon their primary purpose, of truthtelling.
>
> *(2009, p. 131)*

Seen from this perspective, it is not that the *NYT* articles with the *Professional Journalism* frame contained no usable information, but when a news story simply opposes one perspective with another, the readers are left to decide for themselves what the "truth" of the matter is. This was especially true when conflict became the focal point of the stories: the newspaper typically attempted to report all sides, with a common refrain in the articles being that "The company behind the pipeline, Energy Transfer Partners, did not respond to a request for comment." As a result, the *claimsmaking style* used often by the newspaper – within the context of "impartial" professional journalism – served to complicate the ability of the paper to provide a strong environmental justice message.

The other factor complicating the paper's environmental justice framing of #NODAPL is the use of language and imagery that evoked an image of Indigenous protesters as primitive or anachronistic. One article on November 16 titled, "Squash, Rice, and Roadkill: Feeding the Fighters of Standing Rock," focused on the "cutthroat" method of cooks who try to feed many people at the camp. And while the article itself provided a clear vision of day-to-day life at the camp (with an accompanying video interview with the chef), the title and the focus (donated buffalo and roadkill) evokes a kind of primitiveness that stood out among the articles. While many articles focused on justice through the use of Indigenous voices and perspectives, then, the focus on balance and the use of anachronistic stereotypes of "Indians" complicated this.

Despite some significant issues, overall the paper's coverage of the Dakota Access Pipeline protests did not contain *counterrhetorical strategies* that might have denied that a problem existed or blocked calls to action. In addition, the newspaper fulfilled most tenets of the *democratic theory of the media*. Discussed in more detail in Chapter Two, this is a normative theory arguing that the press has several responsibilities to the public. First, in switching from an economic lens to one underscoring the importance of environmental justice for the Standing Rock Sioux, the paper acted as a watchdog, holding the feet of government and pipeline corporations to the fire. The newspaper also attempted to tell the "truth" about the protests, even while some articles languished in "balance." Finally, in its focus on justice for Indigenous groups, the paper appeared to choose a side: according to McChesney (2014, p. 31), "if there is a bias in the amount and tenor of coverage, it should be towards those with the least amount of economic and political power." Overall, then the paper fulfilled most of the normative ideals of journalistic coverage in its treatment of the events at Standing Rock.

One last note is in order about the paper's coverage of the events. As described earlier, the paper's coverage fell off quickly after the vast majority of participants left the various camps associated with the pipeline resistance. Even during some key court cases (notably in June and October, 2017), the paper remained mostly silent. Coverage was mostly limited to a mention of the pipeline in relation to Trump, to a spill from the Keystone pipeline in South Dakota, or in relation to other social and environmental justice issues.

Bismarck Tribune framing analysis

The *Bismarck Tribune* is a daily newspaper that notes on its webpage that it is the official newspaper of North Dakota and the city of Bismarck. Lee Enterprises, which owns the paper, observes some notable moments in its history, including that "a reporter from our newspaper in Bismarck, North Dakota, died with George Custer at the Battle of the Little Big Horn."[14] The paper is a member of the *Associated Press*.

Similar to the coverage by the *New York Times*, the tenor of the articles from the *Bismarck Tribune* showed a shift over time, where early coverage framed the pipeline in a way markedly different from later coverage at the height of the struggle in late 2016. This is where the similarities between the newspapers ends, however, for the *Bismarck Tribune* (*BT*) covered the issue much more frequently, publishing 425 articles from 2015 to early 2018. While the most common frames are described, some frames were only seen a few times, and not all of the less common frames are described at length here.

Because the frame frequency and number shifted during different phases of construction of (and protests against) the pipeline, analysis considers the frames during four distinct phases. Before describing these phases in detail, this section first identifies the most common frames seen throughout the coverage as a precursor to a more detailed discussion of frame frequency and number. Like the *NYT*, the *BT* articles contained *Conflict* and *Professional Journalism* frames; unlike the *NYT*, this newspaper primarily adopted *Law and Order* frames during pipeline construction, and *Economic* and *Law and Order* almost exclusively after construction was over and the oil began to flow through the pipeline.

The frame seen throughout almost all phases of the paper's coverage was *Professional Journalism*. Defined earlier, the focus of this frame remains consistently on "balance" – providing both sides of an issue with minimal context. In this case, the newspaper provided numerous sides – including the Tribe, the protesters, local citizens, law enforcement, the state, and (when it could reach them) pipeline representatives. However, it is important to note that, despite the multiplicity of perspectives, the typical scenario presented the views of the demonstrators as juxtaposed against a panoply of other entities – e.g., the sheriff's office, local citizen groups, the pipeline corporation, and the state), giving the impression of little support from the community for the Tribe and its cause. Articles containing this frame typically were structured in a "he said, she said" format that gave little factual information, only perspectives. As with the *NYT*, this frame tended to "muddy the waters" when it came to framing analysis, for if these articles contained the potential for environmental justice perspective (quotes from Tribal representatives or other organizers), they were consistently countered with an opposing side. Articles like these were categorized as *Professional Journalism*.

Early coverage: May to June 2015

The newspaper started covering the pipeline in late May of 2015, and all four articles during this period exhibited the *David and Goliath* frame. Articles with this frame place primary emphasis on the vulnerable "little guy" whose rights are being trampled upon by powerful corporations or government – in this case, the rights of white landowners impacted by the construction of pipelines. An article on May 31 titled "A New Pipeline Twist" emphasized that the liens placed on the local property owners (due to pipeline construction disputes) were creating stress and "pipeline fatigue." Through interviews with "upset" and "afraid" local landowners, the article quotes one's lament that "It just adds another part of the misery of living in the oil patch." The article mentions the Dakota Access Pipeline by name, noting that one owner would have that running through her land:

> Jennifer Sorenson said it's a complex situation for all of them, because, at least financially, the oil boom has been good for them. "We were all doing fine, but struggling, so these are pennies from heaven," she said. "All of our grandparents homesteaded here. Sometimes I feel like we're selling our soul, our virgin ground, and yes, we're getting money for it."

The landowner's mixed emotions about the pipeline – that is a much-needed "boom" as well as a compromise – are reflected in the article's sympathetic tone when describing landowners' concerns about pipeline intrusions on their land. Images associated with the article include local landowners (wearing denim, checkered shirts, and baseball caps) meeting at a kitchen table or standing on their property.

Another article on June 15 titled "Landowners Leery of Dakota Access Pipeline" described a public hearing in a North Dakota town. The article quotes a man who said he and other landowners did not simply have pipeline fatigue, but instead "We have everything fatigue. Will I sign the easement? Eventually, I'll have no choice." The newspaper was careful to note – several times – that most of the landowners did not have a problem with the pipeline itself, only how the lead up to construction was being handled. Images associated with the article showed Dakota Access Pipeline personnel, some smiling over maps of the pipeline route, and others talking at the public hearing. Also included were large photos of protesting homeowners. A final article on June 17 holds the pipeline company accountable for the way it handled negotiations with landowners by quoting a local father and son: "Howdy Lawler said negotiations should be a two-way street. Instead, it's 'take it or leave it,' he told the commission. 'They don't give a rip about anything but their wallets,' said Rick Lawler."

Articles exhibiting the *David and Goliath* frame support the rights of white landowners, depicting their needs as justified and thus supporting the idea that the pipeline company should be held accountable. The majority of the articles received front-page placement by the paper. In addition, the intimate images (close up, where the landowners make eye contact with the camera) echo the message in these articles: these individuals have important rights that should be upheld.

Leading up to the protests: coverage from March to early May 2016

During this time period there were eight articles. The *Professional Journalism* frame was discerned three times, while half of them (four) employed an *Environmental Justice* frame. The *Economic* frame was seen only once.

Professional Journalism

Once land easements were signed and most negotiations with landowners were complete, the newspaper's coverage demonstrates a shift in framing. In early 2016, one article noted simply that the majority of landowners had signed easements for the pipeline in a just-the-facts tone. Other articles started to focus more attention on the concerns of the Standing Rock Sioux, but still adopted a "balanced tone" that is the hallmark of professional journalism. The article "Pipeline Crossing Raises Concerns," published March 30, encompasses concerns from many sides, including pipeline officials and environmental organizations as well as the Tribe. The article quotes Kelly Morgan, archaeologist with the Tribe, as saying: "We can live without oil, but we cannot live without our water. We know that pipelines break, we know there's spills." The article initially seemed to demonstrate "balance," in the sense that numerous perspectives were directly juxtaposed with the paper itself seeming to take no sides. However, the story also contained numerous reassurances that the pipeline would be safe, noting that

> The corps required one pipeline operator to conduct $1 million of geotechnical borings across the lake and evaluate other routes before it would consider allowing the company to use other methods than directional drilling. Even the Dakota Access Pipeline had to prove its drilling concept by doing borings on both sides of the Missouri River.

These reassurances seemed to undercut the statements by the Tribe, thus exhibiting what Ibarra and Kitsuse (1993) term a *perspectivizing sympathetic counterrhetoric* – that is, while the paper does not deny that the Tribe is concerned, because so many safety precautions were being taken by government and corporations, the paper implies that the Tribe need not worry (and the pipeline construction thus should continue).

Environmental Justice

Other articles during this period, however, did contain *Environmental Justice* framing that focused on the need to consider the Tribe. On March 30, the article "Sioux Spirit Camp to Protest Dakota Access Pipeline," took the perspective and voice of the Tribe and other participants in the pipeline resistance, in the process opening a discussion about sovereignty and nationhood:

> Tribal members, including Chairman Dave Archambault, have taken their concerns about the pipeline's proximity to the corps and the federal Environmental Protection Agency. Steve Sitting Bear, director of external affairs for Standing Rock, said the tribe is pushing for an expanded environmental assessment, since the corps' draft assessment for the crossings on federal lands does not mention Standing Rock. "It's within 1,000 feet of the reservation, but it completely ignores the existence of a tribal nation," said Sitting Bear of the pipeline. "We're hoping to get the information out

there that a tribal nation is put at risk for the interests of big oil and the state of North Dakota – everybody's interests but ours. We'll be the ones in harm's way if this thing breaks."

Other articles in April took this stance, providing in-depth discussions around the Tribe's concerns about environmental safety with their water supply without providing countering perspectives. These pieces demonstrated that the paper took care to give voice to the Tribe during this short period: in reading them there is a general impression that the Tribe's concerns were justified and deserved to be heard. In this sense, this coverage is marked by a concern for justice for the Tribe. The newspaper also included a piece (which was not included in this analysis because it was written by a different news service) giving credence to the idea that the pipeline was originally planned to be placed directly upstream from Bismarck's water supply until local citizens voiced concerns; then it was moved to be directly upstream from the Standing Rock reservation (Dalrymple, 2016).

Economic

As noted, there was only one article during this period that employed economic framing. It is highlighted here because it appears to show the community anticipation of the economic boon from the flowing pipelines as well as the pipeline construction. On May 4, in "Dakota Access Pipeline Reception Strong in Linton," numerous small business owners are quoted: "The business is definitely welcome," the "owner of a gallery and coffee house" exclaimed. The article noted that "between 400 and 700 workers will come through during construction." The reason this article is included here is the way it minimized the growing resistance to the pipeline. At the top of the story was an image of "a small contingent" of Standing Rock tribal members talking to law enforcement. Unlike the earlier images of white landowners, where the visual portraits were intimate and warm, the image topping the article showed the face of a tribal member, eyes closed. The image seems to convey passivity and does not invite the reader to make a meaningful connection to the people associated with the Standing Rock cause.

At the height of the protests: late May 2016 to early March 2017

While for a few months the newspaper portrayed the Dakota Access Pipeline through the lens of environmental justice, the tenor of coverage changed when construction of the pipeline began in late May. This shift in framing continued until President Obama ordered the U.S. Army Corps to review and halt construction in December and the majority of out-of-town protesters started to leave. The primary frames during this period (in 320 articles) were *Law and Order* (129), *Professional Journalism* (82), *Economic* (34), and, less frequently, *Environmental Justice* (26), *State as Environmental Steward* (10) and *Pipelines are Safe* (8).[15]

The *Professional Journalism* frame, well defined and exemplified elsewhere, is not described in detail to make more room for discussion of other frames.

Law and Order

While this frame was rarely visible in the *NYT*, it was frequently employed by the *BT* after Labor Day at the height of the protests, a marked difference between the papers. The majority of the articles demonstrating this frame made the front page (A, p. 1). While numerous examples of this framing exist, a handful of pieces provide particularly vivid illustrations. On October 23, the article "Saturday Protests Lead to 83 Arrests" quoted law enforcement frequently and focused on the "criminal" actions of the protesters:

> Two officers received minor injuries, according to a press release. "This is being termed a riot today because the individuals know they are criminally trespassing," said Capt. Bryan Niewind, of the North Dakota Highway Patrol. "They are creating chaos for law enforcement. They are creating a dangerous environment for us when you have people attaching themselves to equipment." Charges against those arrested include criminal trespass and engaging in a riot.

The word choices in this piece are quite telling, as *riot, chaos, criminal trespass,* and *dangerous* suggests malicious intent as well as disorganization on behalf of the protesters. In addition, the term "riot" (versus "protest") to describe those fighting to protect clean water for the Tribe has a specific and significant connotation. Linguistics professor Nic Subtirelu offers a brief analysis of the significance of the word choice when it comes to framing:

> *Protest* suggests a group of people with a political message, a group of people who are fighting (not necessarily in the literal physically violent sense) for something. Calling something a *protest* almost demands that we mention *what* they are protesting. *Riot* suggests a group of people engaged in senseless, undirected violence. We don't need to hear the political message of rioters, because they have none.
>
> *(2015,* emphasis in original*)*

The newspaper echoed law enforcement's use of the word "riot" many times in its coverage, including an article in mid-October about the use of the word itself. While the article questions the use of such language, it does so within the framework of "balance," and so the reader is left to decide.

The theme of "protesters as criminals" continued throughout the months of active protest. In early November, one *BT* article focused on the possibility that the protesters might have engaged in felonious activities, noting that "A woman accused of firing a handgun at police officers as protesters were evicted from private property... was charged with attempted murder. Red Fawn Fallis faces the most serious charge of any Dakota Access Pipeline protester arrested so far." Images associated with the article included two people with brown skin:

the man was shown in court, wearing an orange prison jumpsuit and unsmiling, while the photo of the woman was a police mug shot showing her in an orange jumpsuit with downcast eyes. The negative characterization of the protesters continued: about a month later, the paper quoted an elected state representative about his characterization of the protests:

> Rep. Kevin Cramer, R–N.D., minced no words about the obstacles in completing the pipeline. 'Roads, bridges, transmission lines, pipelines, wind farms and water lines will be very difficult, if not impossible, to build when criminal behavior is rewarded this way,' said Cramer, who met with Trump on Monday.

In "mincing no words," the paper suggests that Cramer is a straight talker who says it like it is. And, once again, those protesting the pipeline, including the Standing Rock Sioux Tribe, are described through the lens of criminality. Along these lines, in late September, the editorial board made the curious statement that law enforcement had been "patient" with the protesters, who were only arrested because "they wanted to be." Given this perspective, perhaps the title of the editorial was unsurprising: "Tribe Doing Better Job in PR Battle." For the editors of the newspaper, the fight for clean water was reduced to a public relations battle (instead of an environmental justice struggle), and what the North Dakota government, law enforcement, and the pipeline company had been doing wrong was simply not playing to public sentiment well enough.

As noted, the *Law and Order* frame not only defines those resisting the pipeline as violent, criminal, and irrational, but also demonstrates consistent support for law enforcement. In mid-October, during widespread criticism of the tactics used by heavily armed law enforcement, the paper rather curiously referred to this use of force as "proactive policing." This unusual turn of phrase seems to suggest that police's use of force was beneficial because it precluded more violence (as opposed to fueling more of it).

There were numerous examples of positive stories about law enforcement. In early October, as article titled "Backing the Authorities" focused on "a grassroots show of support for law enforcement, particularly during the protests against the Dakota Access Pipeline." In mid-November, another article noted that hundreds of locals had gathered to support law enforcement, including the quote from a local representative that the sheriff's office was "working to protect all of us, not just the farmers and the ranchers and the small town people that live in the area of the protest." And on January 1, 2017, the newspaper gave Sheriff Kirchmeier its award for the 2016 year, noting that

> Law enforcement has worked overtime monitoring protest activities against the four-state, $3.8 billion pipeline project, which has drawn international attention, led to hundreds of arrests and led to millions in law enforcement spending since August. For this, members of the public say they're grateful for public servants as dedicated as Kirchmeier.

The award, noted the paper, "honors members of our community who have gone beyond what's expected." This piece especially instantiates the Law and Order frame for its depiction of both the protesters and law enforcement, for not only does the image associated with the article depict Kirchmeier, but the caption beneath it notes that the sheriff "answers questions on Wednesday after the arrest of eight protesters at the Dakota Access Pipeline construction site" – again drawing attention to the "lawlessness" of the protesters.

The images most often accompanying *Law and Order* were those of law enforcement, and included photos of: the sheriff and other spokespeople giving press conferences; law enforcement vehicles; and law enforcement taking action against protesters. Even when protesters were in the photo, often they were framed by law enforcement. One story provides a representative example: on August 31, when the newspaper reported that "Protesters Disrupt Second Dakota Access Pipeline Worksite," the photo showed protesters who had chained themselves to construction equipment as framed on either side by parked officers' vehicles. Other images in the article included one of James Iron Eyes, Sr. as he sat after being arrested. Fringing him at the top of the photo is a line of police officers. Once again, this visual framing by the *Bismarck Tribune* juxtaposes members of the Standing Rock Sioux Tribe (and their supporters) with images of law enforcement, reinforcing the idea of acrimonious conflict as well as criminality.

Economic

One of the first frames observed during this time frame was *Economic*, which coincided with the start of pipeline construction in May. The placement of stories with this framing tended to be near the front of the paper, and often were on the front page. One early story in April demonstrated this framing by focusing on interviews with local business owners, who stated their enthusiasm for the project because of the additional profits it would bring. By late May, several articles exhibited this focus, including "Pipeline Perks Emmons County," an article on June 12, which noted how many union jobs were being created by the project by quoting a pipeline spokesperson directly.

As the protest activities grew in late summer, the newspaper continued to frame the pipeline in economic terms, but now shifted to how much money was being *lost* due to the protests. Here, the economic frame underscores the financial difficulties faced by the government, the individual taxpayer, local businesses, and the corporation that are due to the Standing Rock protests. On August 25, an article ("Injunction Ruling to Come") gives the following analysis:

> A temporary injunction would have devastating short- and long-term impacts to the DAPL project' the company argued.... Hundreds of deviations from the construction schedule would occur costing as much as $540,000 per occurrence. The aggregate direct impact of these changes would exceed $100 million.

This article (and others like it) highlights the economic burden suffered by Energy Transfer Partners. Others placed the focus on the economic hardship imposed on taxpayers by the Tribe's struggle. A December 1 article ("Line of Credit his $17 M") offers that:

> For a third time, North Dakota Emergency Commission members on Tuesday approved millions of dollars for a line of credit to cover the ongoing costs of law enforcement response to Dakota Access Pipeline protests…bringing the total line of credit to $17 million. Members of the commission used the oft-repeated mantra that they have no choice but to provide for law enforcement and first responders.

In addition to noting the economic strain to the pipeline and the taxpayers, the paper included a third category of financial burden: that placed on local businesses. An article on November 27 identified business owners in nearby towns who were struggling to make ends meet:

> the crowds in Schwieters' store and on the streets were just what local business owners were hoping for, especially after Dakota Access Pipeline protests in Bismarck had made many residents leery. Madonna Wald, owner of Lot 2029, said Nov. 19 was her worst for sales in the six years she's been open in downtown Bismarck. Police had issued alerts warning of protester activity… so sales were down.

The image with this story showed white shoppers walking in downtown Bismarck on a sunny day. This type of economic framing, which creates the sense of significant hardship on multiple levels (pipeline company, taxpayers, and local business owners) places the responsibility for this difficulty onto the Standing Rock Sioux Tribe. This represents an effective *unsympathetic counterrhetoric*, because the paper seems to compare the relative difficulties experienced by different parties and seems to say that the Tribe's concerns are not worth the financial burden placed on the largely white citizenry of Bismarck.

Environmental Justice

The placement of these articles ranged widely, from the front page (A1) back to section B. Although fairly rare in the coverage, when articles with this frame did occur, they were unambiguously oriented towards justice for the Tribe, including their words and perspectives often. A late summer article titled "Pipeline Protesters Vow to Remain" quotes the Tribe and other Indigenous partners at length and does not offer opposing viewpoints:

> "Until this is stopped, a lot of us are dedicated to being here on the ground," said Goldtooth, who called for North Dakota to end its strong-arm tactics,

including air surveillance and a barricade placed on North Dakota Highway 1806.… Young said the tribe has filed a racial discrimination complaint with the United Nations, based upon an earlier decision to locate the pipeline less than a mile upriver from the reservation rather than north of Bismarck. Standing Rock tribal members are calling for the injunction, maintaining that the U.S. Army Corps of Engineers failed to conduct a proper environmental assessment of the water crossing.

Here, the article highlights criticism of local law enforcement's tactics in dealing with those resisting the pipeline and acknowledges that the pipeline was originally set to be placed upstream of the town of Bismarck. In so doing, it invites empathy for those involved in the struggle through personal stories and the Tribe's perspectives. Other articles with this frame during this period paint a detailed portrait of the experiences of the Tribe and those fighting with them. Their positions were not ridiculed, their voices were made clear, some background was provided about treaty history, and the Tribe's cause was portrayed as justified.

Images associated with this frame can be best described as a "mixed bag," where some show law enforcement, others show the protesters, including a close-up of then-Tribal Chairman Archambault. Thus, while *Law and Order* mainly depicted law enforcement, the images associated with the *Environmental Justice* frame seem to have no discernible pattern.

State as Environmental Steward

This frame emerged at the tail end of this time period in January and February, 2017. Winter had settled upon the area, the large-scale resistance to the pipeline had paused due to the U.S. Army Corps (temporarily) denying a key permit. The focus was now on what would happen to many of the camps, including cleanup of the large amount of refuse. This frame, then, positions the state of North Dakota as a responsible environmental caretaker that looks after the camp once the majority of people had left. Within this positioning is an obverse definition of those involved in #NODAPL, including the Tribe, as irresponsible, childlike (one enduring stereotype of Indigenous peoples[16]), and even hypocritical: while they have fought against an oil pipeline, this frame suggests that it is they who now endanger the river with camp waste.

An article in mid-February titled "More Resources Applied to the Cleanup" demonstrates this frame by quoting the state's attorney general admonishing the protesters, noting that while he supports free speech, he will use law enforcement to "do what is necessary to protect public health and safety. It's time for protesters to either go home, or move to a legal site where they can peaceably continue their activities without risk of further harm to the environment." Later, the article cites Burgum's policy adviser Bachmeier: "We want to prevent trouble and protect the environment. Our focus is to clean up the camp and make sure the river is not polluted." An article earlier in February also quoted Governor Burgum's administration regarding the state's desire to clean up the protest camp: "We're all on the clean water team." On February 4 there was more talk of government cleanup, this time including the U.S. Army Corps' perspective:

> "As stewards of the public lands and natural resources, we have a responsibility to the public to prevent injuries and loss of life, and to ensure that our precious water resources are free from pollution due to human activities and respect for all who rely on this water for their livelihoods," said Col. John Henderson, the corps' Omaha District commander.

The majority of these stories were located on the front page of the newspaper. Images associated with many of the articles were aerial views of the floodplain on which the camp stood, noting how the garbage left behind would soon enter the waterways once the snow melted and the rivers flooded.

Pipelines Are Safe

This frame occurred rarely during coverage, which, out of hundreds of articles, might not be worth including in a summary of analysis. Two statements from the newspapers' editorial board, however, used this frame primarily, and thus it is worth acknowledging in analysis. In late August, the BT editorial board wrote a piece exhorting "all sides" to be patient. While the board initially appeared to support no side in particular, a paragraph in the statement was significant:

> If there's a loser, it's Dakota Access Pipeline. While work continues on the project in many other areas, it's been stopped in Morton County. So there are financial losses for the company. The Tribune Editorial Board also feels a reminder is in order. The pipeline project comes with many safeguards including a monitoring and shutoff system. While there's no 100 percent guarantee with any form of transportation, this is a state-of- the-art system.

In this statement, the pipeline is presented briefly in economic terms (how much money has been lost) but the central point is that pipelines are safe because the industry is trustworthy and has created "safeguards." The use of "state-of-the-art" phrasing is telling, because it implies that the pipeline corporation has the newest and safest technology available. Early in the next month, the editorial board wrote another piece ("Issues Run Deeper Than the Pipeline"), one seemingly more sympathetic to the Tribe at first because it acknowledged their concerns about environmental justice. But then the board wrote:

> The Tribune Editorial Board, early during the oil boom, endorsed the use of a network of pipelines to move oil. We felt then, and still do, that pipelines provide the most efficient and safe means for moving oil. Pipelines aren't accident-proof, but new pipelines have more safeguards and monitoring systems than those built in the past. This one will be thicker than normal in

the water crossing, bored 90 feet below the bed of the river and the crossing pipe "pigged" – that's an interior inspection device – at least every five years.

Both of the editorials represent what Ibarra and Kitsuse (1993) refer to as the *perspectivizing sympathetic counterrhetoric*, where a claimant's stated need is portrayed as one perspective, not necessarily as an objective "truth." In this case, the "claimant" is the Tribe, which expresses a fear over their water supply. While the editorial does not deny that the Tribe has concerns, it then states (as fact) that their fears are unfounded due to the inherent safety of pipelines. What this ultimately means, then, is that the paper defines resistance to the pipeline as having little justification.

After the demonstrations (March 2017–March 2018)

This final phase of coverage (which at the time of this writing is ongoing) places almost exclusive focus on two aspects of the pipeline and the resulting protests: numbers demonstrating the economic benefit from the oil flowing from the Bakken fields to external markets, and detailed chronicling of the court cases arising from the protests. There were 95 articles published during this time. The paper's coverage is thus distinct from the other newspapers, which had largely ended their coverage in February of 2017. It is perhaps not surprising that the paper showed no signs of abating as the paper is local, and local profits and prominent trials would feature in this type of coverage. What is notable, however, is that *Law and Order* (49) and *Economic* (13) frames were the most common lenses through which the public encountered the Tribe and other protesters during this time. Economic stories tended to quantify the amount of oil moving through the pipelines and note how much it would mean to the economy. Less frequent frames included *Empathy* (7) which often manifested as a sympathetic perspective on the struggles faced by the Tribe; *State as Tribal Friend* (4) involved positioning the new governor Doug Burgum as the initiator of reconciliation towards the Tribe; and *Corporate Malfeasance* (3) which identifies corporate wrongdoing in relation to violations caught by the state.[17]

Law and Order

This frame, as noted, was featured in almost half of the stories during this time after the demonstrations had died down. Like the articles assigned to this frame in previous phases, these focused on who had been arrested during the protests; unlike previous *Law and Order* stories, these were focused almost exclusively on what crimes allegedly had been committed, what court decisions had been made, and summary statistics of crimes and convictions. Almost without exception, these stories were located in the "Courts and Crime" section of the paper. Images associated with these articles varied, but often included mug shots of the accused or depicted the accused in court. Sometimes the images would include a defendant inside their home, outside the courtroom, or marching in protest but, importantly, these were most often reserved for white demonstrators.

During this phase, the voice of the Tribe was almost completely missing. Then-tribal chairman Dave Archambault III was quoted in two articles, and other tribal members were included in ten articles. This was a significant change from previous phases of coverage, when the Tribe was heard quite often, even if it was directly opposed by the state or law enforcement. The voices heard most often during this time period were those relating to court, most often including judges and lawyers.

It is important to note that some of the difficulty involving the identification of articles with the *Law and Order* frame involved both content and placement. That is, while a few of the articles took a professional approach that cited all sides, the articles would still be focused primarily on the alleged criminal acts of the protesters and were placed in the paper's "Crime and Courts" section, ultimately aligning with the frame of *Law and Order*. Articles like these were analyzed carefully, but most ended up being identified as *Law and Order* rather than a *Professional Journalism* frame. One reason for this is provided by Stuart Hall and colleagues in their well-known "Policing the Crisis" (1978) book, where they identify typical elements of a law and order approach: police statements regarding particular cases; crime reports (usually with statistics and perceived trends); and stories solely focused on court cases (p. 69). The stories in the newspaper fell within these three types, and thus were classified as *Law and Order*.

One article demonstrating the trends and statistics of arrests was on December 29. Titled "Dakota Access Cases Could Be Closed in 2018," this story presented readers with a bulleted list of court case decisions, noting that on December 28, "499 protest cases had closed, with 229 remaining open, 98 inactive with warrants and three on appeal." Another article a few weeks later on January 16 ("Bound for Trials") placed focus on Red Dawn Fallis, who was

> set to be tried in two weeks as the first federal defendant in connection to the Dakota Access Pipeline protests. She is accused of firing a handgun at officers as they arrested her for disorderly behavior, but her case isn't the only one in federal court. Assistant U.S. Attorney Gary Delorme confirmed six other defendants are indicted in connection to the months long pipeline protests.

Although it notes other court cases, the newspaper placed particular focus on Fallis in particular throughout this phase: she was mentioned quite often (in eight articles), likely due to the alleged attack on law enforcement and the federal nature of the offenses. Almost always, Fallis was depicted, unsmiling, with unkempt hair, in a mug shot.

One article in early 2017 stands out within the *Law and Order* framing. On March 29, an article titled "Chase Iron Eyes Pleads Not Guilty" notes that

> Iron Eyes, 39, is charged along with 28-year-old Vanessa Castle with directing people to set up camp across from the main Oceti Sakowin protest camp on Feb. 1. The land, which some protesters saw as rightfully theirs by treaty,

is owned by the Dakota Access Pipeline company.... At issue in the case is whether the day-long demonstration, in which pipeline protesters set up teepees in a new "Last Child Camp," was a riot and what role Iron Eyes played.

The passage is intriguing because while it reveals the perspective of the protesters – in that they perceived that they had rights relating to historical treaties – the paper does not attempt to discern the truth but instead reduces their views to simply a misconception held by the "water protectors" – a type of *perspectivizing counterrhetoric* that the paper used often to subtly negate the needs of the Tribe.

As noted in the earlier definition of *Law and Order* for this research, the flip side of the coin for this frame includes a positive focus on law enforcement. Stories like these included: praise for law enforcement's role in pushing back the #NODAPL protests, personal pieces on individual officers, and condemnation of those who sought to portray law enforcement in a negative light. An editorial on April 26, 2017, exhibits several of these elements. Titled "Courts Show System Works for Everyone," the paper's editorial board contends that

> During the protests, pipeline opponents tried to tarnish the reputation of law enforcement, Morton County and the state by arguing tactics and arrests went beyond the norm. The court results so far don't support that argument. *While many of the misdemeanors have been dismissed, it doesn't mean law enforcement went too far. Many of those charged wanted to be arrested as part of the effort to disrupt the pipeline work.* The courts decided in some cases that those arrested weren't given proper notice through sign postings or warnings from personnel that they were on private land.... That's how the legal system should work. Law enforcement officers used their best judgment when making arrests and the courts decided whether they followed proper procedures. *Those who had their cases dismissed may feel vindicated, but they also should be pleased the system works as intended.* [emphasis added]

The article is notable in several regards. First, it praises the tactics taken by law enforcement, arguing that they were necessary for pipeline work to continue. Then, it posits that the majority of protesters wanted to be arrested (which would include a desire for fines and jail time). Finally, the board suggests that those who were arrested should show appreciation for the legal system, despite the numerous stories that had emerged indicating that the vast majority of protesters could not afford the time or money required for court appearances. This idea promoted by the board – the legal system is fair and works for all concerned – was seen a few times in their editorials. On June 4, 2017, in "Protest Cases Show System That Works," the board contended that the acquittal of then-tribal chairman Archambault provides "another example of how the system has worked for pipeline opponents" without acknowledgment of the alternative perspective – namely that the "system" had arrested an Indigenous leader for protesting an oil pipeline being placed upstream from his tribe's water supply in the first place.

Summary of Bismarck Tribune *coverage*

This framing analysis proved to be complex, because while some frames persisted throughout coverage, other frames came and went as the resistance and pipeline construction went through different phases. While the *Environmental Justice* frame was uncommon when considered throughout the entire span of the newspaper's coverage, the paper was fairly sympathetic to the concerns of the Tribe early on. However, while the violent events of Labor Day appeared to precipitate a pivot towards justice framing at the *New York Times*, those same events seemed to harden the *Bismarck Tribune* against the Tribe's plight, manifested in its significant shift to a *Law and Order* framing.

When considering the valence of the frames in terms of empathy for the Tribe throughout the coverage, there were many more articles that presented the Tribe in a negative light (as a draw on resources, as violent lawbreakers, as water contaminators) than positive (water protectors as peaceful with legitimate environmental concerns). The placement of the articles also was intriguing: stories with "negative" frames like *Law and Order* consistently made it onto the front page (A, p. 1) of the paper, whereas stories that framed the protesters in a positive light made it onto the front page far less often. Put another way: when the "water protectors" were peacefully protesting or made a positive contribution to community, they were frequently shunted to the back pages, whereas news of them being violent or trespassing were more often given front-page treatment.

Finally, another word about "balance" is in order. While the protesters (Indigenous or otherwise) were quoted in many articles, their perspective was often directly opposed by representatives from the pipeline corporation, the state of North Dakota, local law enforcement, or local citizens. In so doing, the Tribe was always fighting to have their perspective acknowledged as important or based in some form of truth. In this sense, the pieces from the paper's editorial board were particularly revealing, in that while they consistently acknowledged the Tribe's environmental concerns, they made clear that they believed that pipelines were the safest means to transport oil, thus consistently creating a *perspectivizing sympathetic counterrhetoric* that ultimately negated the concerns of the Tribe. Analysis reveals, then, that the newspaper did not fulfill the normative ideals of the *democratic theory of media*. Although they were the media source closest to the conflict, the continued reliance on a "he said, she said" format rendered the truth elusive, for who (if not a local paper with boots on the ground) would confirm for readers what had actually happened? In addition, the paper did not privilege the perspectives of those with historically fewer rights but instead favored law enforcement and the state. Finally, it did not act as a watchdog and did not hold Energy Transfer Partners or the state accountable for placing the pipeline directly upstream from the water supply of the Standing Rock reservation.

While a journalistic focus on "balance" tends to preclude much-needed context for an issue (and this was true for most of the paper's coverage), the paper did provide one article in November that spoke clearly to treaty rights.

Titled "Treaty Deepens the DAPL Discussion," the thoughtful, in-depth piece described the Treaty of Fort Laramie of 1851 as possibly providing rights to the Tribe. Interviews were given with historians and Native American advocates making just this case. Because the treaty was treated as being legitimate and significant, this article provided a sense of the Tribe being justified in its fight. Although this lone article was categorized as *Environmental Justice*, the very next day the paper resumed its *Law and Order* framing of the pipeline battle.

Examples like this make the paper appear to be conflicted, and the numerous frames speak to that. In addition, the paper did not engage meaningfully with the environmental justice perspective used by the #NODAPL movement – and thus the Tribe appeared to have little influence on the newspaper's framing of the issue. Interestingly, there was another recurring phenomenon, where many articles had what this research would categorize as *Law and Order* titles (e.g., how many protesters were arrested for criminal activity) but whose actual article content reflected a different frame. These took time to analyze, but the impression that emerges is of a paper that felt somewhat divided and torn about its treatment of the Standing Rock Sioux Tribe.

Indian Country Media Network framing analysis

According to its website, *Indian Country Media Network (ICMN)* is "the world's largest and most trusted news source for contemporary Native American and indigenous news." According to Google Analytics, the website had "registered 1,009,761 unique monthly visitors for the month of June 2014." The website was created in early 2011 and has won several awards from the respected *Native Americans Journalists Association*.[18] The frames exhibited the most frequently in *ICMN*'s 46 stories about Standing Rock are *Environmental Justice* (22) and *Empathy* (12); far less frequently the news stories contained *Law Enforcement Overreach* (3), *Professional Journalism* (3), *Political Perspective* (3), and *Call to Action* (3).

Environmental Justice

According to the organization's website, coverage of the Dakota Access Pipeline began on March 19, 2016, with an article titled "Dakota Access Pipeline Threat: What You Need to Know." The story notes that while the pipeline corporation claims that all necessary precautions have been taken to ensure safety,

> dozens of environmental organizations, individual landowners, concerned residents and one tribal government strongly disagree. The potential for an oil spill is always a risk with a pipeline project, they say, and if that were to happen, the environmental consequences would be devastating.

Here, the article uses two important words – "threat" and "devastating" – that evoke a sense of danger for those on the Standing Rock Reservation. The image associated with the article was the Missouri River, highlighting that the environmental threat involves the significant pollution of water from the oil pipeline. The article then notes that

> While the pipeline would not technically run directly through the Standing Rock Reservation, it would cross the Missouri River only a few hundred feet upstream from Standing Rock's border…comprising the entire eastern border of the reservation. If a leak were to occur, it would undoubtedly devastate the environment, people, resources and land of the Standing Rock nation. The quality of the water of the Missouri River is critical to the health and well-being of the tribe, both economically and culturally.

Here, the word "devastate" is used again, and the environmental needs of the Tribe are underscored. This article fell under the category of the *Environmental Justice* frame. Defined earlier, the central components of this frame include a focus on the environmental needs of the Tribe with a recognition of the need for justice, especially involving historical and contemporary cases of environmental racism. Other stories demonstrating this frame placed even more emphasis on the history of environmental injustice in relation to Indigenous groups in the U.S. A few months later on August 20, a story titled "Standing Rock Sioux Issue Urgent Appeal to United Nations Human Rights Officials" framed the pipeline construction as a human rights issue by citing the *U.N. Declaration on the Rights of Indigenous Peoples* in relation to the Standing Rock Sioux Tribe's appeal, including the

> "right to health, right to water and subsistence, threats against sacred sites including burial grounds, Treaty Rights, cultural and ceremonial practices, free prior and informed consent, traditional lands and resources including water, productive capacity of the environment, and self-determination," the appeal said. It cites environmental racism stemming from the Army Corp's decision not to locate the pipeline north of Bismarck over concerns it would endanger the city's water supply, while issuing permits to trench through burial grounds and the Tribe's main water supply.

The idea presented here – that the pipeline construction represents "environmental racism" – is mentioned directly in the interpretation of the United Nations' declaration, especially in regard to the decision to move the pipeline after concerns were expressed from the majority-white Bismarck. This view continues with the recognition of the disturbance of ground considered sacred by the Tribe. Regarding the disturbance of sites considered sacred by the Tribe, one article on September 13 described the pipeline construction thusly: "Dakota Access oil pipeline workers gouged a trench over two miles of Sioux burial grounds on September 3 near Cannon Ball." The term "gouge" is particularly evocative, for the Cambridge dictionary defines the verb as "to make a hole in something in a rough or violent way." Gouging the land thus seems to be a way for the journalist to describe the pain and sense of violation experienced by the Tribe during construction.

Images associated with the *Environmental Justice* frame included: members of the Tribe and other people who had joined the struggle, the Missouri River and (less frequently) site construction and police versus protesters. The heavy preponderance of Indigenous faces juxtaposed with the river seems to tell a visual story regarding the Tribe's struggle for clean water. It also echoes the unswerving focus on the Indigenous perspective in this conflict, which leads to a discussion of the second most common frame: *Empathy*.

Empathy

In his discussion of framing, Schudson (2011, p. 40) argues that, "A female reporter is more likely than a male reporter, other things being equal, to see rape as a newsworthy issue. An African-American reporter is more likely than a white reporter…to find issues in the African-American community newsworthy." *Empathy* was the second most common frame seen in *ICMN*. The Cambridge dictionary defines empathy as "the ability to share someone else's feelings or experiences," and the *ICMN* outlet demonstrated this ability numerous times in about one-quarter of all articles. It certainly could be argued that the *Environmental Justice* frame contains a component of empathy, and yet there were many articles that clearly expressed empathy without referencing treaties or a history of injustice based on racial identity. And because the journalists who write for *ICMN* are Native American (from various nations across the U.S.), empathy also makes sense, for many of these journalists have lived experiences relating to the issues faced by the Standing Rock Sioux Tribe.

The *Empathy* frame was seen in the numerous stories that focused predominantly on the perspectives of the Tribe. On February 2, 2017, a story ("Temporary Restraining Order of Dakota Access Pipeline Construction Denied") quotes Cheyenne River Sioux Tribal Chairman Harold Frazier:

> Frazier expressed disappointment in the ruling that denied a temporary restraining order. "It was pretty tough," he said. "One thing that we do know is that our water is our life, it is life," he continued, "…and any damage to that water will completely destroy us entirely."

Similarly, a September 17, 2016, article ("Dakota Access Lake Oahe Work Stopped Pending Standing Rock Sioux Appeal") used the words of then-Standing Rock Sioux Chairman Archambault: "This is a temporary administrative injunction and is meant to maintain status quo while the court decides what to do with the Tribe's motion," he said in a statement. "The Tribe appreciates this brief reprieve from pipeline construction and will continue to oppose this project."

What makes stories like these demonstrate an empathetic framing is not simply quoting Tribal members – the *New York Times* and *Bismarck Tribune* did that at length. It was the fact that their voices were allowed to stand as a form of truth, without simply being directly and immediately opposed by another, contradicting viewpoint. In addition, *ICMN* adopted the language used by the Standing Rock Sioux Tribe and allies to identify themselves – "water protectors," demonstrating the empathetic stance.

Additional frames

The small number of stories with the *Law Enforcement Overreach* frame contained references to a militarized police force and often used examples of prayerful and unarmed "water protectors" being arrested. Articles containing *Professional Journalism* cited (or tried to cite) all sides, including Energy Transfer Partners. Stories exhibiting the *Political Perspective* frame focused on Obama or other politicians who supported the Tribe during the struggle. Finally, the *Call to Action* framing provided readers with specific actions they could take – including online activities, donating money or supplies, talking to elected representatives, and withdrawing money from banks that supported the pipeline project.

In addition to containing frames that were different in type or frequency than that seen in the other two newspapers, there were certain stories that *ICMN* picked up that did not get "play" in the more mainstream news outlets. First, the events of Labor Day, when the bloody clash occurred between protesters and pipeline security guards, was covered in detail by the site. An article titled "'And Then the Dogs Came': Dakota Access Gets Violent, Destroys Graves, Sacred Sites," provided multiple quotes from the "water defenders" on scene that chronicled the violence, including references to "snarling dogs," people being "flipped," and construction workers using their personal trucks "as weapons." Another story picked up by *ICMN* was in reference to the Indigenous youth media movement "Rezpect Our Waters." This story was covered by the outlet numerous times, but unmentioned by the other U.S. newspapers.

Summary of ICMN coverage

In all, the *ICMN* news outlet covered the pipeline construction and #NODAPL movement through the lens of *Environmental justice* and *Empathy*, not permitting the professional goal of "balance" deter it from providing the perspective and voice of the Tribe and its allies in the struggle against the pipeline. While the goal of professional journalism is unswerving commitment to detachment and impartiality, there is another way for journalism, one more focused on what Valeria Alia (2004, p. 94) terms "thoughtful subjectivity," one that gives voice to those with "less power and often more information about a range of constituencies and perspectives."[19] In other words, subjectivity, when done with intention, can "empower the disempowered" (94).

Returning, for a final time, to the *democratic theory of media*, it is clear that the outlet fulfilled the tenets of this normative theory by not only acting as a watchdog to those in power but also giving voice to "those with the least amount of economic and political power" (McChesney, 2014, p. 13). Simply put, *ICMN* covered stories that were not covered by the mainstream press and, in all but a few cases, did not attempt to "balance out" the Tribe's perspectives. It provided deep historical context for injustices faced by not only the Standing Rock Sioux Tribe but for many other Native American groups.

Summary: comparison of U.S. news sources

As the analysis has revealed, the frames employed by each news outlet differed by type and frequency. The *Indian Country Media Network* provided much-needed historical context for the issue in addition to giving primacy to Indigenous voices and perspectives during the struggle. The *New York Times* eventually adopted an environmental justice framing, losing some of its adherence to the professional standard of "balance" for a cause it thought was justified. While the *Bismarck Tribune* at times used the lens of environmental justice, it consistently employed the language and rhetoric of "law and order," where the protesters were portrayed as a threat to the social fabric. As Boykoff (2006, p. 227) observes, the stakes are relatively high for those hoping to gain attention for their cause:

> Mass-media attention is crucial to social movement development. Yet even if social movements are able to work their way under the "media spotlight," as Wisler and Giugni put it, they may receive mass-media coverage that could do them more harm than good. The news media—through framing practices—set the parameters of acceptable public discourse. Voices that fall outside the range of acceptable discourse are occasionally permitted space on the mass-mediatized terrain, but their price of admission is often subjection to mass-media deprecation.

The reasons for the differing coverage cannot be known definitively, but it seems likely that location made a difference. That is, the *New York Times* was somewhat removed from the issue – enough so that it could justify condemning police actions when its board felt they had gone too far. It is not to say that the paper did not have any financial interest in the project (either through advertising revenue streams or other sources), but none was readily apparent. The *Bismarck Tribune*, however, is tied to a state and region that is deeply connected to fossil fuels. North Dakota is a state whose economy experienced a revival with not only the fracking that allowed Bakken fossil fuels to be harvested, but also the means to transport that oil out of state. When Stoddart (2007, p. 665) analyzed the *Vancouver Sun*, he found that: "the environmental discourse that enters the Sun's coverage of environmental policy debate is ultimately compatible with the ongoing dominance of a capitalist political economy." The same could be said, truthfully, of both papers (Chapter One makes that case for all professional news organizations), but it appears the *New York Times* was willing to break ranks for what it perceived to be injustice, and the *Bismarck Tribune* was not.

The *Indian Country Media Network* is an unusual case in this regard: while the message on its website reveals that it needs to reconsider its funding model, there were no glossy ads for cars, shoes, or oil companies surrounding the articles on #NODAPL. In other words, there seemed to be no conflict of interest in terms of corporate interest or revenue streams. And yet, journalism like this that exhibits empathy and a concern for environmental justice clearly struggles in a type of paradox: if outlets like *ICMN* don't take corporate money, they avoid the corporate pressure to report on important issues through the professional journalism lens of balance but they then struggle to stay afloat and reach a wider audience. If they *do* take corporate money, they will likely lose their important public service function that provides a much-needed counterbalance to profit-driven, commercial media. As Schudson recognizes,

> Empathy, fortunately for all of us, is not beyond human capacity, and good journalism may both stem from empathy and evoke it. Still, the person who writes the story matters. When minorities and women and people who have known poverty or misfortune firsthand are both authors of news and its readers, the social world represented in the news expands and changes.
>
> *(2011, p. 40)*

Notes

1 For many of these observations, Larson cites Murphy and Murphy (1981), *Let My People Know*, and Weston (1996), *Native Americans in the News*.
2 One example of this is the "Indians on the Warpath" trope seen in news coverage of past conflicts, including the takeover of the Wounded Knee site by the American Indian Movement in 1973 (Loew, 2011).
3 Seymour notes that there are three central problems with using "panoramic" terms to describe Indigenous groups: 1) the term is as useful and specific as the use of the word "European" to describe someone from France or Norway; 2) overuse of the term enables a view of Indigenous, sovereign nations as lacking distinction or difference; 3) "race-based overgeneralizations" have been linked to "pan-tribal frames" (p. 77).
4 McChesney cites Christians et al. (2008) for an expanded version of this theory.
5 Noted earlier, McChesney here has expanded upon his earlier version (McChesney, 2008, p. 25) of a democratic theory of media that does not include, in the earlier version, the third tenet presented here.
6 Snow and Benford (1992) note that the use of frames in academia is marked by interdisciplinarity, including psychology, sociology, the humanities, and more.
7 Gamson and Modigliani (1989) make a distinction between "framing devices" (i.e., those that propose a way to consider the issue) and "reasoning devices" (i.e., those that suggest how the issue should be addressed) (p. 3).
8 Gamson and Modigliani (1989) further specify different types of reasoning devices: 1) "roots" (referring to analysis of the causes of the problem); 2) "consequences" (considering effects and impacts); and 3) "appeals to principle" (regarding moral claims made about the consequences).
9 According to Ibarra and Kitsuse (1993), there are four different types of "sympathetic" counterrhetorics (that recognize any given issue as a significant "problem" but reject the need to address it): 1) *Naturalising*, used to justify the status quo in terms of inevitablity and/or as a natural state of being; 2) *Costs Involved*, where the costs of addressing the problem are seen to be too high; 3) *Declaring Impotence*, which points to "an impoverishment of resources at hand for dealing with the issue" (p. 39); and 4) *Perspectivizing*, where a claimant's stated need is portrayed as one perspective, not necessarily as an objective "truth." All of these accept "claimants' complaint...while withholding support for remedial action" (p. 39). Unsympathetic counterrhetorics, on the other hand, "oppose the condition-category's candidacy as a social problem and therefore also reject the call for remedial activities" (p. 41) through four different approaches: 1) *Antipatternings* occur when the claimant's problem is depicted as an isolated issue and not a larger pattern; 2) *The Telling Anecdote*, where one example is used to disprove the claim of a wider

problem; 3) *Insincerity* suggests a "hidden agenda" on the part of the claimant (p. 41); *Hysteria* suggests an irrationality or emotionality to a group's claims, thereby negating their claims to a true problem.

10 This statement comes from the editorial board description on the NYT webpage at http://www.nytimes.com/interactive/opinion/editorialboard.html

11 An early, abbreviated version of this research was presented with my graduate student Erica Tucker at the *Conference on Communication and the Environment* in Leicester, UK, 2017. During this initial stage of research, we examined fewer articles from the *New York Times* and *Bismarck Tribune*, did not consider the *Indian Country Media Network*, and did not include images in analysis. The majority of frames identified in the early research are used here, including *Economic, Professional Journalism, Law and Order,* and *Conflict.* The abstract for this talk is available at https://theieca.org/sites/default/files/COCE_2017_program/_program.html

12 The NYT Annual Report is available at http://s1.q4cdn.com/156149269/files/doc_financials/annual/2016/Final-Web-Ready-Bookmarked-Annual-Report-(1).pdf

13 Walker (2011) also notes that the U.S. environmental justice frame includes the geographical impact of environmental racism outside of strict national borders, and that "other versions of an environmental justice frame have emerged within the US government and its agencies." He also discusses the EPA's use of this framing (p. 23).

14 According to Erdoes (1982, p. 18), the first newspaper to report on the news of Custer's death was the *Bismarck Tribune* on July 6, 1876,

15 Approximately 31 articles were uncategorized due to ambiguity where a frame could not be easily discerned.

16 Per Larson (2006, p. 109).

17 As before, for some articles it was difficult to determine a frame and thus approximately 19 of these remain uncategorized.

18 This information currently is available at https://indiancountrymedianetwork.com/education/native-education/indian-country-today-media-network-reaches-online-traffic-milestone/

19 Alia (2004) cites scholar Rita Shelton Deverell as a key advocate of subjectivity in journalism.

References

Alanis-Ramirez, K., & Tena, B. (2017, July 1). Examining the advocacy campaigns for #NoDAPL and how it can inform future campaigns like the Comanche Trail pipeline protest. Paper presented at the *Conference on Communication and the Environment*, University of Leicester, June 29–July 2.

Alia, V. (1999). *Un/covering the north news, media, and aboriginal people.* Vancouver, BC: UBC Press.

Alia, V. (2004). *Media ethics and social change.* New York: Routledge.

Barthel, M. (2017). Despite subscription surges for largest U.S. newspapers, circulation and revenue fall for industry overall. *ACI Information Group.* Retrieved from http://scholar.aci.info/view/14bd17773a1000e0009/15c649f787f0001a860d3d4

Boykoff, J. (2006). Framing dissent: Mass-media coverage of the global justice movement. *New Political Science, 28*(2), 201–228. 10.1080/07393140600679967

Christians, C. G., Fackler, M., Richardson, K., Kreshel, P., & Woods, R. (2008). *Media ethics: Cases and moral reasoning* (8th ed.). Boston, MA: Allyn & Bacon.

Dalrymple, A. (2016, August 18). Pipeline route plan first called for crossing north of Bismarck. *Bismarck Tribune.* Retrieved from https://bismarcktribune.com/news/state-and-regional/pipeline-route-plan-first-called-for-crossing-north-of-bismarck/article_64d053e4-8a1a-5198-a1dd-498d386c933c.html

David, C. C., Atun, J. M., Fille, E., & Monterola, C. (2011). Finding frames: Comparing two methods of frame analysis. *Communication Methods and Measures, 5*(4), 329–351. 10.1080/19312458.2011.624873

Davies, N. (2009). *Flat earth news*. London: Vintage Books.

Deacon, L., Baxter, J., & Buzzelli, M. (2015). Environmental justice: An exploratory snapshot through the lens of Canada's mainstream news media. *Canadian Geographer / Le Géographe Canadien, 59*(4), 419–432. 10.1111/cag.12223

Entman, R. M. (1993). Framing: Toward clarification of a fractured paradigm. *Journal of Communication, 43*(4), 51–58. 10.1111/j.1460-2466.1993.tb01304.x

Faber, D. (2008). *Capitalizing on environmental injustice: The polluter-industrial complex in the age of globalization*. Lanham, MD: Rowman & Littlefield.

Falc, E. (2017, July 2). "Water protectors": The framing of environmental collective action at standing rock. Paper presented at the *Conference on Communication and the Environment*, University of Leicester, June 29–July 2.

Gamson, W., & Modigliani, A. (1987). The changing culture of affirmative action. *Research in Political Sociology. A Research Annual, 3*, 137–177. ISSN 0895-9935.

Gamson, W., & Modigliani, A. (1989). Media discourse and public opinion on nuclear power: A constructionist approach. *American Journal of Sociology, 95*(1), 1–37.

Gitlin, T. (1980). *The whole world is watching*. Berkeley, CA: University of California Press.

Hall, S. Critcher, C., Jefferson, T., Clarke, J. & Roberts, B. (1978). *Policing the crisis: Mugging, the state, and law and order*. New York: Holmes & Meier.

Haluza-Delay, R. (2007). Environmental justice in Canada. *Local Environment, 12*(6), 557–564. 10.1080/13549830701657323.

Hansen, A. (2010). *Environment, media and communication*. London: Routledge.

Hopke, J. E. (2012). Water gives life: Framing an environmental justice movement in the mainstream and alternative salvadoran press. *Environmental Communication: A Journal of Nature and Culture, 6*(3), 365–382. 10.1080/17524032.2012.695742.

Ibarra, P., & Kitsuse, J. (1993). Vernacular constituents of moral discourse: An interactionist proposal for the study of social problems. In G. Miller & J. Holstein (Eds.), *Constructionist controversies* (pp. 21–54). New York: Walter de Gruyter.

Klyde-Silverstein, L. (2012). The 'Fighting whites' phenomenon: An interpretive analysis of media coverage of an American Indian mascot issue. In M. Carstarphen & J. Sanchez (Eds.), *American Indians and the mass media* (pp. 113–127). Norman, OK: University of Oklahoma Press.

Lakoff, G. (2010). Why it matters how we frame the environment. *Environmental Communication: A Journal of Nature and Culture, 4*(1), 70–81. 10.1080/1752403090 3529749.

Larson, S. G. (2006). *Media & minorities: The politics of race in news and entertainment*. Lanham, MD: Rowman & Littlefield.

Loew, P. (2011). Finding a new voice – foundations for American Indian media. In M. Carstarphen & J. Sanchez (Eds.), *American Indians and the mass media* (pp. 3–6). Norman, OK: University of Oklahoma Press.

Matthes, J. & Kohring, M. (2008). The content analysis of media frames: Toward improving reliability and validity. *Journal of Communication, 58*(2), 258–279. 10.1111/j.1460-2466.2008.00384.x.

McChesney, R. (2014). The struggle for democratic media: Lessons from the north and from the left. In C. Martens, E. Vivares, & R. McChesney (Eds.), *The international political economy of communication: Media and power in South America* (pp. 11–30). London: Palgrave Macmillan.

Moore, E. E., & Lanthorn, K. R. (2017). Framing disaster. *Journal of Communication Inquiry, 41*(3), 227–249. 10.1177/0196859917706348

Pellow, D. N., & Brulle, R. J. (2005). *Power, justice, and the environment: A critical appraisal of the environmental justice movement.* Cambridge, MA: MIT.

Phillips, S. (2012). "Indians on our warpath": World War II images of American Indians in *Life* magazine, 1937–1949. In M. Carstarphen & J. Sanchez (Eds.), *American Indians and the mass media* (pp. 33–55). Norman, OK: University of Oklahoma Press.

Sanchez, J. (2012). American Indian news frames in America's first newspaper, *Publick Occurrences both Foreign and Domestick.* In M. Carstarphen & J. Sanchez (Eds.), *American Indians and the mass media* (1st ed.). (pp. 9–17). Norman, OK: University of Oklahoma Press.

Scheufele, D. A. (1999). Framing as a theory of media effects. *Journal of Communication, 49*(1), 103–122. 10.1111/j.1460-2466.1999.tb02784.x

Schudson, M. (2011). *The sociology of news* (2nd ed.). New York: W.W. Norton & Company.

Seymour, R. (2012). Names, not nations: Patterned references to indigenous Americans in the New York Times and Los Angeles times, 1999–2000. In M. Carstarphen & J. Sanchez (Eds.), *American Indians and the mass media* (pp. 73–93). Oklahoma City: University of Oklahoma Press.

Snow, D., & Benford, R. (1992). Master frames and cycles of protest. *Frontiers in Social Movement Theory.* New Haven, CT: Yale University Press.

Stoddart, M. C. J. (2007). 'British Columbia is open for business': Environmental justice and working forest news in the Vancouver Sun. *Local Environment, 12*(6), 663–674. 10.1080/13549830701664113.

Subtirelu, N. (2015). Covering Baltimore: Protest or riot? Retrieved from http://scholar.aci.info/view/1456b4242ee14f10104/14d05fb99c80001000e

Tallent, R. (2013, May 1). Don't misrepresent Native Americans. *The Quill, 101,* 26.

Valkenburg, P. M., & Semetko, H. A. (2000). Framing European politics: A content analysis of press and television news. *Journal of Communication, 50*(2), 93–109. 10.1111/j.1460-2466.2000.tb02843.x.

Vickery, J., & Hunter, L. M. (2016). Native Americans: Where in environmental justice research? *Society & Natural Resources, 29*(1), 36–52. 10.1080/08941920.2015.1045644.

Walker, G. (2011). *Environmental justice: Concepts, evidence and politics.* Hoboken, NJ: Taylor & Francis.

Weston, M. A. (1996). *Native Americans in the news: Images of Indians in the twentieth century press.* Westport, CO: Greenwood Press.

5

"COULD IT HAPPEN HERE?"

Canadian newspaper framing of the Dakota Access Pipeline

The previous chapter provided the definition of framing as used in this research and presented the results of framing analysis for coverage of resistance to the Dakota Access Pipeline by three news outlets in the U.S., finding key differences in the type and frequency of frames used by each source. This chapter builds from that analysis and provides a point of comparison through framing analysis of three newspapers in Canada: the *Calgary Herald*, the *National Post*, and the *Globe and Mail*. The end of this chapter contains two types of summaries: one comparing results between Canadian newspapers (with a special focus on the widespread omission of the Dakota Access Pipeline resistance), and one broadly comparing the differences in coverage of the issue between U.S. and Canadian news outlets.

A *frame* was defined in the last chapter as a "central organizing idea or storyline" (per Gamson & Modigliani, 1989, p. 3). Because framing has been defined at length in Chapter Four, it need not be revisited here except to briefly note the focal points for this analysis. Frames are seen to reveal moral judgments about the cause and scope of a given problem, along with ideas about potential solutions and responsible parties – in other words, they "*define problems...diagnose causes...make moral judgments...and suggest remedies*" (Entman, 1993, p. 52, emphasis in original). Frames also can pose the question of "what problem?" through rhetorical devices like *sympathetic counterrhetorics* (per Ibarra & Kitsuse, 1993), thereby negating the idea that there is an issue that requires solving. As such, frames reflect "the play of power and boundaries of discourse over an issue" (Entman, 1993, p. 55), and this is certainly true of the wave of resistance that grew in opposition to a major pipeline being constructed directly upstream from the Standing Rock Sioux reservation.

Based on earlier, seminal work on social constructionism and framing (e.g., Gamson & Modigliani, 1989; Entman, 1993; Ibarra & Kitsuse, 1993), analysis here focuses on: 1) the imagery published with the stories (when available); 2) the specific language and rhetoric used in the articles, including particular phrases and slogans as well as adjectives used to describe the actions and characters of protesters, police, the local public, and government; 3) prominent and recurring examples used in the stories; 4) the tone and style of the articles; and 5) the sources quoted most commonly. This analysis permits a detailed portrayal of how each news outlet chose to portray the resistance that met the construction of the pipeline. In addition, it is important to note that while framing is treated as both a method and a theory, framing as an interpretive framework is complemented by normative theories such as the *Democratic Theory of Media*, which underscores the need for a socially responsible press with a strong public service function through truth telling and holding the powerful accountable. As such, this chapter ends with a discussion of the implications of the framing analysis results within these normative theoretical frameworks.

Selection of newspapers for analysis

All Canadian newspapers selected for analysis were located using the same methods described in Chapter Four (searching *Nexis Uni* and *Access World News* with the keywords "Dakota Access Pipeline"). Based on the search of the two databases, I discovered that the two national papers covering the Dakota Access Pipeline the most were the *Globe and Mail* and the *National Post*. The provincial Alberta papers that covered the issue the most were the *Calgary Herald* and the *Calgary Sun*, with the former containing slightly more articles than the latter. The fact that a small provincial paper covered the issue at all is likely due to the fact that Alberta has numerous active oil and gas fields, is home to many energy companies, and also must consider Indigenous issues and rights when it comes to oil development and transportation.

It is important to note at the outset that identifying Canadian newspapers that covered the events surrounding the Dakota Access Pipeline was a significant and often frustrating challenge. In part this seems to be due to the widespread news media blackout on Standing Rock – a silence covered in depth by some advocacy organizations and news outlets, including Fairness and Accuracy in Reporting, Minnesota Public Radio, and Al Jazeera America as well as by well-known Indigenous journalists like Eugene Tapahe and Mark Trahant. Although the media blackout was generally understood to be a global phenomenon, the Canadian dailies were particularly silent on the demonstrations.

Complicating the relative paucity of articles available for analysis from Canadian newspapers was that while many dailies I initially selected – including the *Toronto Star* and the *Chronicle Herald* in Halifax, Nova Scotia – initially seemed to have a sufficient number of articles, a deeper inspection revealed that many articles were authored by other freelance news organizations, including (frequently) the *Associated Press* and *Canadian Press*. For example, while a database search for *Toronto Star* yielded 57 articles, the majority of these did not discuss the pipeline directly or came in the form of reprints from the *Washington Post*; and while the *Chronicle Herald* had published 58 articles, additional inspection revealed that stories reprinted from the *Associated Press* or *Canadian Press* constituted the vast majority (roughly 50) of all articles published by the paper.[1]

Significantly, when a Canadian paper *did* author stories on the pipeline resistance, almost every article centered on the implicit question, "could it happen here?" As such, while the #NODAPL movement was in fact covered (albeit lightly) by several Canadian dailies, the events at Standing Rock simply seemed to be a lens through which to view issues occurring north of the U.S. border. This meant that the focus was less on the movement as a standalone event worthy of consideration and more on how it would impact the Candian "oilpatch" and its related industries – a theme I return to later.

As a result of these myriad challenges, a case could be made for excluding Canadian news coverage from framing analysis. However, traditional framing analysis and scholarship would consider the relatively low number of articles published by Canadian newspapers an important finding in itself, and one worthy of exploration through the lens of omission. In addition, I chose the papers that covered the issue the most throughout the country, maximizing the sample size available for analysis. I also found clear patterns within each paper but also between them, despite the relatively low number of articles. Finally, the fact that both the *Bismarck Tribune* and *Calgary Herald* are regional newspapers located in heavy oil-producing areas made them an interesting subject of study and worth comparing to the larger, more national dailies in each country.

Framing Analysis of the *Calgary Herald*

The *Calgary Herald* is a daily newspaper that first opened in 1883[2] with 62,974 weekday circulation, according to News Media Canada (2016).[3] It is owned by *Postmedia Network* (*PN*), which owns one daily – the *National Post* – and numerous provincial newspapers. In Calgary itself, *PN* also owns the *Calgary Sun*. According to the *Canadian Encyclopedia*,

> From the early days the Calgary Herald was a supporter of the pioneer ranching industry; this support has since expanded to include the area's oil and gas interests. The paper is often at odds with the views expressed in other parts of the country, and even with Alberta's other major newspaper, the Edmonton Journal. National awards have been won in news, photography and production.[4]

Calgary Herald *frames*

Due to exclusion in this research of articles not written by the newspaper itself, the number of articles (16) from the *Calgary Herald* is relatively small, making omission an important lens through which to view this coverage. The paper did not cover the origins of the resistance movement in April, 2016, nor did it publish stories on the widely publicized events of Labor Day, 2016, when security personnel used dogs on Indigenous demonstrators. Thus, despite the similarities between Bismarck and Calgary (both are located within heavy oil-producing regions that are interested in extending their pipeline network, and both are located in regions with Indigenous populations), the paper chose to cover the issue relatively infrequently. Despite this, however, clear patterns emerged in the coverage with two common frames. The frame of *Law and Order* (which portrays protesters as dangerous, unpredictable, and irrational while primarily taking the perspective of law enforcement[5]) was discernible in six articles. The *Economic* frame (which defines the pipeline and protests only through the lens of money – from speculating about the profits to be made as well as potential losses suffered if pipelines are not built) was seen eight times, or in half of the articles. Other frames seen only once included *Protests are Contagious* (which expressed fears that the Standing Rock demonstrations would move north to protest Canadian pipelines) and *Professional Journalism*.

Economic

For this commonly seen frame it is important to note that most articles written by the *Calgary Herald*, regardless of frame, expressed an underlying, consistent concern – namely, "could it happen here?" This is interesting, because it reveals that the pipeline protests interested the paper only insofar as they might impact the oil and gas industry in Calgary, in Alberta and, more broadly, in Canada. One article in particular stands out for its use of language to describe the protesters through an economic lens. In "Activists Put Heat on Enbridge to Address First Nations Rights," the story first provides a quote from a member of the Standing Rock Sioux Tribe – one of the only instances where the Tribe's perspective and voice was included. The article then quickly moved to a discussion of the need for the energy transportation company Enbridge to disclose how it planned to incorporate the possibility of Indigenous protest into its business plan for pipelines:

> Monaco said the company would disclose in future reports how it handles indigenous issues when executing projects, its indigenous policies, and its efforts to consult early and often on projects. He later said during the heated protests over Dakota Access that attracted thousands of people last year, Enbridge had reconsidered its investment in the project, which had cleared many regulatory hurdles by the time the company invested.... Monaco said the company assessed whether previous consultations with indigenous communities were adequate and whether project proponents had responded to what they had heard.

Another article ("Non-profit Applies New Pressure on Enbridge; Indigenous Rights, Environment the Focus of Resolution"), takes a critical perspective of Enbridge, noting that the corporation was facing fresh calls to explain how it assesses indigenous rights and environmental risks "facing new investments after the Dakota Access Pipeline ignited a firestorm south of the border." The image accompanying this depicted protesters speaking at a rally; underneath the photo the caption read "Enbridge is being asked to disclose the weight it gives to indigenous and environmental concerns when making new acquisitions." From these two examples it is clear that even when the Tribe was given a voice in the paper, it was framed not through the lens of environmental justice or human rights but instead through economics.

Another example of the *Economic* frame comes shortly after the election of Trump on December 14 in an article published in the business section titled "Trump's Energy Triumvirate: Big Implications for Alberta." This frame was visible through references to some of President Trump's appointees (specifically, Rex Tillerson to U.S. Secretary of State, Scott Pruitt to head the Environmental Protection Agency, and Rick Perry as Secretary of Energy) that took on a political tone:

> It's hard for Canada's energy sector, treated like a poor cousin under the Obama administration, not to feel giddy about the abrupt change. So, are good times about to roll again for Alberta's oilpatch? The short answer is not yet. There are still too many variables and it takes time for a resurgence in investment to take root. That said, things look much better than they did 12 months ago. For that, everyone is breathing a cautious sigh of relief.

The article places emphasis on the importance of investment in Alberta's oil and gas industry, noting that the American pro-business president Trump would be encouraging a "resurgence" that would be a much-needed "relief" for "everyone." By "everyone," the article seems to imply that the benefits of having a resurgence in oil and gas infrastructure would apply to all.

The enthusiasm for the economic benefit of pipelines was seen in numerous articles. Omitted was any concern regarding the fate of the Standing Rock Sioux Tribe or for any First Nations' people north of the U.S. border. The frame as expressed in Canada is best seen in an article in late January titled "Trump Revives Keystone Pipeline; 'Very Good Moment for Alberta' Could Bring Jobs, Higher Oil Price." The article title is perhaps self-explanatory, but of note is the celebratory tone: noting that Trump had revived both the Keystone XL and Dakota Access Pipeline, the article includes a quote from Canadian Natural Resources Minister Jim Carr, who called Trump's moves

> a very good moment for Alberta... This decision will result in many, many jobs for Albertans and it's also a sign that there is a recognition by the new American administration that Canada can be a source of economic development and of job creation on both sides of the border.

Law and Order

The second most frequent frame in the newspapers' coverage of the North Dakota pipeline construction and protest was *Law and Order*. The first article displaying this frame was published in mid-November, 2016, shortly after the U.S. presidential election, which noted that Trump had promised to pave the way for more oil and gas infrastructure projects in the U.S.:

> The controversial Dakota Access Pipeline is also likely to be a beneficiary. The $3.8-billion line has faced escalating opposition from environmental groups and has been stalled by the administration's pending review. Friday, law officers arrested about three dozen Dakota Access protesters in a confrontation that also shut down a state highway. The midday incident began after about 100 protesters confronted crews doing dirt work along the pipeline route where pipe had already been laid.

This article follows a similar format for the *Law and Order* frame, where protesters are depicted as the aggressor. In this case, the article suggests that they "confronted" construction crews unnecessarily, because the pipeline work was complete already. In addition, the story notes how many protesters were arrested. Intriguingly, the article does not identify that the protesters included the Standing Rock Sioux Tribe, only (rather oddly) that they were members of "environmental groups," thus leaving out the potential for an environmental justice element to the story.

Other articles struck a similar tone that, like the *Economic* framing described above, had a "could it happen here?" flavor. Stories like these tended to note the arrests in North Dakota as a sign that the "criminality" could cross the border into Canada. One article, which noted the "celebration" of the oil industry over the approval of the Enbridge pipeline, wrote that

> Opponents have threatened a repeat of the 2016–2017 protests on the Standing Rock Reservation in North Dakota against the Dakota Access pipeline, in which Enbridge owns a stake. Those protests drew thousands of opponents and there were more than 700 arrests.

National Post analysis

Circulation (weekday) for this news outlet in 2016 was over 183,000 (including both paid and unpaid subscriptions), according to News Media Canada.[6] Similar to the *Calgary Herald*, the paper started in the late 1990s by Conrad Black, who had purchased the *Financial Post* and subsumed that paper into the *National Post*, with the *Financial Post* being retained as the business section of the central paper. The paper has struggled financially, being owned by Black, then CanWest Global, and finally Postmedia Network, which owns it currently. In addition, the paper has curtailed its circulation in the eastern provinces and publishes less often in the move to a more online format. Potter (2014) notes that "The National Post competes directly with the Globe and Mail, and is considered to have a more conservative stance."

As with the *Calgary Herald*, the selection of articles was done carefully. While the general criterion for article selection for all newspapers (including the U.S. and Canada) in this study was that opinion pieces would be excluded, I noticed an interesting trend for the *National Post*, where those identified as journalists or editors would be published in the news site but also in the paper's "Comment" section. In this section, the paper notes, one could "read through unique Editorials and Business & Finance columns from our top editors covering current events." In Chapter Four I noted that while opinion pieces from non-journalists were not considered, editorials expressing the views of the newspapers' editorial staff were included, especially because, in the words of the *New York Times*, they most clearly reflect "the voice of the board, its editor, and the publisher." Thus, those articles were cautiously included to be consistent with the selection criteria. Of note is that this paper seemed to have many more of these types of editorials than any other newspaper, and with tones more strikingly opinionated than editorials from the *Bismarck Tribune* or the *New York Times*. As a result, there is little ambiguity regarding the position of the paper or its staff. Articles written by outside organizations were still excluded, in keeping with the selection process for all newspapers.

National Post *frames*

Similar to the *Calgary Herald*, once duplicates and articles written by other news sources were excluded, the number of articles (14) available to be analyzed was relatively small. As with the *Calgary Herald*, however, very clear patterns of coverage were discernible even in this relatively small number of news stories, and thus were included in the research. There were two primary frames that emerged from analysis: *Economic* (seen the most frequently – seven times) and *Law and Order* (in six articles). Seen only once was the frame of *Professional Journalism*.

Economic framing

Similar to other newspapers in the U.S. and Canada, the *Economic* frame seen in this paper still defined the Dakota Access Pipeline – and the resistance that grew around it – through the lens of profits to be made or lost. Articles with this frame included a January 25, 2017, article titled "Trump Endorses Oil, Embraces Pipelines," which framed the U.S. President's approval of various pipelines as good for the economy:

> U.S. President Donald Trump delivered on his promise to approve the Keystone XL pipeline, signing an executive order Tuesday to advance its construction and giving Canada's battered oil sands industry his support. "We are going to renegotiate some of the terms" of the Keystone XL project, Trump said to reporters. "And if they like (them) we will see if we can get that pipeline built – a lot of jobs, 28,000 jobs, great construction jobs." He took the same action on the Dakota Access pipeline project, saying that it would be "subject to terms and conditions negotiated by us."

The idea of fulfilling a "promise" strikes a positive tone, especially through the idea that thousands of jobs will be created as a result of pipeline approval. The striking language of a "battered" oil industry also suggests that the oil industry is the David to environmental organization's Goliath, and thus represents the more vulnerable party. This inclusion is especially interesting given the article's omission: and although it notes the impending Dakota Access Pipeline approval, it does not mention the Indigenous struggles over it. At the top of the article was a slideshow of images that included President Trump signing the order about the pipelines, pipeline construction and, very briefly, some protesters of other pipelines. The slideshow appears to reinforce the *Economic* frame in the sense that Trump is directly juxtaposed with imagery of jobs created from construction.

On June 26, 2017 ("Trump Delivering on His Promises"), an editorial notes Trump's accomplishments when it comes to the economy, taking a conservative political tone (in addition to an economic framing):

> President Trump's detractors consider him a buffoon who can't get anything done. They do so at their peril. In just five months in office, Trump has racked up a jaw-dropping string of accomplishments. This "buffoon" is buffaloing his critics, most of whom are too blind to see how effectively he's implementing his agenda. *The Keystone and Dakota Access pipelines are no longer stymied. Business is buoyed by the prospect of lower taxes and less regulation, leading to business confidence and to a stock market that's at record highs.* [emphasis added]

The photo at the top of the article shows Trump at one of his rallies, arms spread while talking, the U.S. Presidential seal beneath him. This article, like the one described above, underscores Trump's authority and his commitment to economic growth by highlighting the stock market and the buoying of business and industry. The Dakota Access Pipeline, in this view, is just another step towards more jobs and more economic strength.

Although the articles highlighted above are fairly typical for the *Economic* frame as defined in this research, the *National Post* in particular included another element within this frame – namely that environmental concerns, because they may conflict with economic growth, are unjustified and nonsensical.

Law and Order framing

The second most common frame distinguishable in the *National Post* articles was *Law and Order*. An October 25, 2017, article titled "U.S. Terror Law to Be Tested on Pipeline Activism" instantiates this frame well. This article describes efforts by various U.S. lawmakers to discern

> whether the domestic terrorism law would cover actions by protesters that shut oil pipelines last year, a move that could potentially increase political rhetoric against climate change activists.... The move by the lawmakers is

a sign of increasing tensions between activists protesting projects including Energy Transfer Partners LP's Dakota Access Pipeline and the administration of President Donald Trump, which is seeking to make the country "energy dominant" by boosting domestic oil, gas and coal output.

Here, the article suggests that the Dakota Access Pipeline protesters may be "terrorists" as defined by law, juxtaposing this idea with the notion that these protesters are stymying President Trump's attempt to bolster the oil and gas industry. This, perhaps, represents the height of *Law and Order* framing, where the protesters actions may not simply break the law, but instead could be classified as acts of terrorism that endanger society. This article, by itself, could simply be interpreted as providing readers with the latest information about the events surrounding the pipeline protests; taken in conjunction with the relatively small number of articles, however, it becomes clear that the paper consistently chose a certain lens through which to view the issue: as a threat to law and order in part due to the challenge to the oil industry and economic growth.

Economic and Law and Order frames

A point made in this chapter and the next is that analysis reveals a trend in terms of the two frames seen most often in the *National Post*. One article in December of 2017 provides a particularly good instantiation of the complementary nature of the *Economic* and *Law and Order* frames, "Project Delays Proving Costly," describing the history of pipeline protests and how they might be different now:

> Anti-pipeline activists from these and other groups had devoted years of campaigning to stop all three projects through regulatory, political and legal means, before adopting a more "in your face" approach, in the vein of the ongoing uprising against the now stalled Dakota Access Pipeline in the United States, and in the tradition of the famous War in the Woods two decades ago against logging on Vancouver Island's Clayoquot Sound. Resistance to infrastructure projects has become the norm in Canada's resource sectors. As part of a four month investigation, the Financial Post identified 35 projects worth $129 billion in direct investment – mostly private money – that are struggling to move forward or have been sidelined altogether because of opposition from environmental, aboriginal and/or community groups.

In focusing on the money to be lost from "opposition" to pipelines due to the changing, illegal, "in your face" protest trend, this article weaves together economic concern with the "lawless" nature of the protests as a way to describe the hurdles faced by pipeline corporations and the oil industry in general. It also moves seamlessly between the protests south of the border and what might happen in Canada as a result.

The *Globe and Mail* framing analysis

The *Globe and Mail*, according to Potter et al. (2015), was created in 1936 by George McCullagh when he combined two daily newspapers – *The Globe* and *The Mail and Empire*. Woodbridge Co. Ltd has been the sole owner of the paper since 2015. Its weekday circulation, according to a *News Media Canada* estimate, is almost 347,000.

Globe and Mail *frames: coverage of Dakota Access Pipeline*

The *Globe and Mail* began covering the Dakota Access Pipeline in late August 2016 and stopped coverage in early August 2017 with a total of 22 articles. Fourteen of the articles were published within the business section, which was titled "Report on Business" ("ROB"). As the placement of articles suggests, the majority of the articles were focused on economic concerns regarding all North American pipelines (and not simply the Dakota Access Pipeline), and the majority of the paper's coverage of the Standing Rock protests were included as a way to talk about the potential impact of worldwide protests on Canadian pipelines like Trans Mountain and Keystone XL. However, as will be explained in more depth in the upcoming section, this was the most challenging analysis yet, for while the placement of the articles suggested an economic focus, a thread of empathy ran through almost all of the articles regardless of frame, where Indigenous leaders from various nations were quoted often (in over half of the articles) and their perspective given. Unlike other newspapers (from the U.S. and Canada) included for analysis here, this economic framing by the paper was not paired with any articles that contained the *Law and Order* frame. The frames discerned in the *Globe and Mail's* coverage were, in order of frequency, *Economic* (seen in 12 stories), *Professional Journalism* (6), *Empathy* (2), *Environmental Justice* (1), and *Devil's Bargain* (1).

Economic

As noted, the *Economic* frame was the most common for this paper. As with other stories demonstrating this frame, the focus was placed on the Dakota Access Pipelines through the lens of how much money would be made or lost as a result of the protests, including a focus on jobs and the potential for economic growth (or stagnation) for the country on the whole. This framing, however, differed from other Canadian dailies in the *way* in which an economic framework was applied to the stories. That is, while the focus remained on pipeline companies' stock prices (the stock value of Enbridge, Inc., a pipeline and infrastructure corporation, was frequently posted at the end of these stories) and job creation, Indigenous voices were frequently used (in 11, or half of the articles), which could be interpreted as a counterbalance to the idea that economic growth trumps all other concerns. However, this too is complex, and thus is discussed more in the summary section at the end of this chapter. For now, some examples reveal how the economic framing manifested in the *Globe and Mail*.

On September 12, 2016, early in the paper's coverage of the issue, an article titled "First Nation sets pipeline precedent; Sioux community and allies win a reprieve on proposed expansion of Trans Mountain project" reveals how an economic framing for this paper also includes a "can it happen here?" flavor in the following lengthy passage:

> From across Canada and the United States, aboriginal leaders rallied to the cause of the Sioux community, which argued the proposed pipeline would disrupt burial grounds and threaten the reservation's water supply. The standoff at the Missouri River is the latest hot spot in a continent-wide struggle between indigenous communities and an oil industry eager to build additional pipelines to move crude to global markets, or to carry it from recently developed fields like North Dakota's Bakken. In Canada as well as in the United States, First Nations are asserting the right to approve or reject pipeline projects that traverse their traditional territory and demanding to be treated as full partners by industry sponsors and governments. In Canada as well as in the United States, governments are struggling to recognize the United Nations principle of free, prior and informed consent without handing every First Nation that could be impacted by a resource project a veto of its fate.

In the discussion of the interests and power of Indigenous groups in the U.S. and Canada, the paper suggests major similarities between the two countries: they both have oil, have proposed and active pipelines, and must contend with the power of Indigenous nations who might resist oil and gas infrastructure. There is also an implicit recognition of the power of these Indigenous nations in the reference to the United Nations.

The article leads with a description of Reuben George, a member of the "Tsleil Waututh First Nation," who "celebrated with the aboriginal protesters at the Standing Rock Sioux reservation last week when the Obama administration delayed a \$3.7-billion (U.S.) oil pipeline project that would run past the community straddling the North and South Dakota border." The photo at the top of the article contained an image of George. The story also includes the voice of the Assembly of First Nations (AFN) National Chief Perry Bellegarde: "No pipeline construction should ever begin until... the indigenous peoples [who are impacted] have provided their free, prior and informed consent consistent with the United Nations declaration on the rights of indigenous people." Along these lines, the story provided subtle criticism of Jim Carr, Natural Resources Minister in Canada, by noting that, when it comes to sovereignty and the power of Indigenous groups to hold sway over a project, "he dodged the question of consent." The article, located in the business section (B3), ends with the admonition that "As the Liberal government weighs later this fall whether to approve the expansion of the TransMountain project, ministers will have to consider the precedent of Standing Rock."

Other articles with an *Economic* framing included one on September 23 that made the front of the business section. This story led with a discussion of Standing Rock protests as a way to discuss First Nations' interests and pipelines in Canada, noting that many Tribes in the U.S. and Canada signed a "treaty" in solidarity to stop pipelines:

> Representatives of some U.S. tribes have also signed the accord, as they engage in a high-profile effort at the Standing Rock reservation in North and South Dakota to block construction of the Dakota Access pipeline. That project would carry crude from the prolific Bakken field to Illinois.
>
> "Indigenous people have been standing up together everywhere in the face of new destructive fossil fuel projects, with no better example than at Standing Rock," said Grand Chief Stewart Phillip, president of the Union of BC Indian Chiefs.

The article goes on to note that "not every First Nation leader is against new pipelines," providing examples from other Indigenous leaders stating that they "need to get pipelines built. We certainly need the people to be on board and not be abrasive toward the concept of having pipelines." This inclusion retains the economic framing, but this time employing it to demonstrate a benefit to Indigenous groups who could see financial windfall from the pipeline projects.

The last article containing a reference to the Dakota Access Pipeline (on August 7, 2017), continued the trend of an economic frame that still included the Indigenous perspective. In "Enbridge Starts Line 3 Rebuild, But Protests Won't Stop," numerous details are provided regarding the pipeline corporation's work and associated "capital costs" associated with infrastructure construction, but then ended by quoting the Chippewa Tribe:

> In a submission, the Fond du Lac Band of Lake Superior Chippewa said it has not been persuaded of the need for it, and if it is built, they want it to be far from their territory. "It is worth noting that this proposed project threatens the headwater regions of the two largest watersheds in North America, the St. Lawrence/Great Lakes and the Mississippi," the band said in its submission. "While the religious, spiritual and cultural significance of Lake Superior to the band cannot be overstated, it should be obvious enough to the broader population that the largest freshwater lake in the world must be protected."

This article, like most of the others with an economic framing, was published on the front page of the business section. And while the general content of the story is in regard to financial and logistical information about Enbridge, it then gives the Chippewa the coveted last word, letting the Tribe's words about environmental protection hang in the air.

Professional Journalism

The second most common frame in the *Globe and Mail* was *Professional Journalism*. Stories with this frame tended to be published throughout different sections of the paper, including section A, the Focus section, and Business. A trend seen in the coverage by Canadian papers, these stories contained a thread of "will it happen here?" In this case, the Standing Rock struggle became seen as a canary in the coal mine, of sorts, for Canadian oil and gas infrastructure. In these types of stories, all major perspectives were provided without seeming to privilege any one side. Articles with this frame included the very first article written on the Dakota Access Pipeline on August 26 titled "Dakota Access Opponents Call for Scrutiny," which warned that the Standing Rock struggle sets

> a pipeline precedent: Opponents of the Dakota Access crude pipeline say the domestic U.S. project deserves the same rigorous environmental and regulatory scrutiny as the controversial, Canadian, cross-border Keystone XL. TransCanada Corp.'s proposed Keystone XL – designed to move mainly heavy oil from Alberta – was put under the microscope by U.S. regulators for seven years before it was ultimately rejected by President Barack Obama in late 2015. In contrast, the Sioux-led demonstrators now camped out in a North Dakota field say the $3.78-billion (U.S.) pipeline proposed by Energy Transfer Partners LP was rushed through state and federal approval processes.

The article cites many sides, seeming to privilege none, while consistently providing the Indigenous voice – in this case, that of Standing Rock Sioux Tribe Chairman:

> "Domestic projects of this magnitude should clearly be evaluated in their totality – but without closer scrutiny, the proposal breezed through the four state processes," David Archambault II, chairman of the Standing Rock Sioux tribe, wrote in an opinion piece for the *New York Times* this week.

Retaining the thread of seeming apprehension running through the other articles, the story ends with

> like many other pipeline stories, there is also a Canadian angle to the Dakota Access saga. Calgary-based Enbridge Inc., through its affiliate Enbridge Energy Partners, announced on Aug. 2 that it would acquire a 27.6-percent indirect interest in the Bakken pipeline system that includes the Dakota Access pipeline.

Other articles that contained a professional lens included an October 12 article titled "Climate Activists Force Pipelines to Shut Down: Group Says It Is Acting in Solidarity with the Protesters at the Standing Rock Sioux Reservation." The tone of the article, even while describing the protests, stays neutral, citing Spectra

Energy Corp and Enbridge Inc. as well as activists from the group "Climate Direct Action," who said they took their cue from

> "anti-pipeline protests led by North and South Dakota's Standing Rock Sioux Reservation," but also took the "personal, direct action" to push the U.S. government to enact stricter measures on climate change. The group is calling for a total ban on new fossil fuel extraction and an immediate end to oil sands and coal use. The moves mark a further escalation in the battle against Alberta's oil sands industry as it seeks wider access to global markets.

In articles like these, all sides were given equal space to express concerns, perspectives, and goals but, as in most articles with a *Professional Journalism* frame, they adopted the familiar "he said, she said" format that precluded any other, more meaningful perspectives.

Empathy

As noted, two of the articles from the *Globe and Mail* analyzed for this research included the frame of *Empathy*, which acknowledges the Tribe's concerns as valid and empathizes with their struggle. One article, titled "Bellegarde Emphasizes Importance of Indigenous Consent for Pipelines" and published October 7, creates a clear sense of the need to understand the Indigenous perspective by quoting the AFN Grand Chief:

> In a speech to an industry conference, Mr. Bellegarde said First Nations people in Canada are determined to assert their right to self-determination, and that includes free, prior and informed consent over energy projects that affect their traditional territory. Mr. Bellegarde said opponents to pipelines may mount high-profile demonstrations aimed at blocking construction, as has happened around the Standing Rock Sioux reservation in North and South Dakota, where tribes are opposing a pipeline to take crude from the Bakken shale oil fields to refineries in the Midwest. Direct action "is definitely an option – like Standing Rock in the States," he said after his speech to the Positive Energy conference at the University of Ottawa.

The image associated with the story depicted Bellegarde speaking at the conference in front of a podium, his arms spread wide. In this story, Bellegarde's perspective is not counterbalanced with any other views and thus stands on its own. As with all the other articles from the Globe and Mail, references to Standing Rock and the Dakota Access Pipeline are couched in terms of what it means for Canada. Although this story was published in the business section of the paper, it was categorized as *Empathy* due to its sympathetic tone and lack of counterbalancing perspective.

The other article that demonstrated the *Empathy* frame was published in early December. Titled "Opposition Prepares to Fight Trans Mountain Approval,"

Rueben George is once again included and his Tribal affiliation is provided while noting the similarities to Standing Rock protests. The article notes that "demonstrations" were still ongoing in North Dakota, and it was George's hope that his protest will not "turn violent." The photo accompanying the article again shows a close-up of George's face. This story was published in section A of the paper.

Environmental Justice

There was, as noted above, only one article that exhibited an *Environmental Justice* frame, published in section A of the paper. The story publicized the award given to an Edmonton photographer who captured images of the Standing Rock protests. The article contained numerous images, including one of Vonda Long, whom the articles notes is a

> descendant of High Hawk, who was killed in the 1890 Wounded Knee Massacre. She says she carries trauma from colonization but has been fighting for justice her whole life as a member of the American Indian Movement. "That's what you do. You sacrifice for your brother." Despite the images of struggle, she says she was surrounded by constant prayer during her time in Standing Rock. "Everything was prayer. There was smudging at the gates, food preparation was prayer, interacting with the horses was considered a form of prayer," she said. "Every activity brought people back to a common purpose. They weren't protesting the pipeline, they were praying against it."

Other images included a man, eyes closed, tears streaming from his red face (presumably from tear gas) who was helped by another person who poured bottled water into his eyes. One photo depicted a young boy running into the "Unity Tepee" at the Sacred Stone camp. The words accompanying the images notes that

> Footage of the standoff at Standing Rock over the Dakota Access Pipeline may have been dominated by clouds of tear gas and riot police, but photographer Amber Bracken set out to show the wider context of a community struggling to determine its future.... "The goal was to contextualize this issue. It's not that the Sioux just decided this year that they didn't want the pipeline. It's not about a specific project, it's about their right to determine what happens on their land,"; she said on Monday after winning the award.

The images associated with the article visually highlight what the Standing Rock protesters endured during the months-long conflict, encouraging readers to see the issue as an environmental justice issue integrally tied to racial considerations. However, it is intriguing that even this article cannot escape a tie to Canada, for it was the fact that a Canadian photographer won an award that provided the key thread woven through the story.

Canadian newspaper coverage of Dakota Access Pipeline

There is nothing magical about a frame. What I mean is simply that, although certain scholars may define framing slightly differently, or outline the parameters of a particular frame distinctly for their own research, the underlying question for all framing analysis is "how are news audiences supposed to feel about this issue (or person, or event) after reading this article or story?" In other words, in what way does a news story define: 1) the issue at hand in terms of cause (who or what is to blame); 2) the consequences (who or what is impacted by the issue); and 3) the solution, including who should be involved/responsible. Per Gamson and Modigliani (1987), "a frame generally implies a policy direction or implicit answer to what should be done about the issue" (p. 143).

Key findings and issues

Of course, a significant consideration in all of this is whether a news outlet chooses to define what constitutes a "problem" at all, which prompts the question of how Canadian newspapers either minimized or outright denied the human rights struggle over the pipeline. This happened in a few ways, the first of which being the infrequent coverage by all major dailies except the *Globe and Mail* which, with 22 articles, had the most coverage of all dailies identified in the database search. This number, while "high" relative to other Canadian newspapers, is of course still very low in comparison to the U.S. news outlets included and assessed in the previous chapter.

Second, in addition to not covering the issue frequently, two of the papers included in this research – the *National Post* and the *Calgary Herald* – seemed to ignore one of the key (if not the key) player in the pipeline struggle: the Standing Rock Sioux Tribe. Instead, these papers placed their focus squarely on the actions of large, well-known environmental groups who became allied with the Tribe during the pipeline resistance. The exclusion of the Tribe as an important player is highly significant, for it effectively erases the idea of Indigenous rights and sovereignty that had bolstered the Tribe's case for environmental justice. Thus, in many of the stories from the *Calgary Herald* and *National Post*, the Tribe seems not to exist at all, with various environmental groups standing in as proxy in the fight against the pipeline. As a result, these types of stories serve to reduce the idea of agency exercised by the Tribe as well as simply erasing them from the conversation.

The third way that the idea of a "problem" with the Dakota Access Pipeline was minimized came in the form of shunting the story about the resistance to the back pages and/or the business section of the papers. Noted earlier, the *Globe and Mail* placed the bulk of their articles about the issue in the business section of the paper. Perhaps it is obvious enough not to state explicitly, but placing stories about the pipeline struggle in the business sections of the papers served to frame the issue in terms of economics – and perhaps more specifically in terms of what Beder (1998) terms a "cost-benefit analysis" where environmental concern always loses out to economic growth and profit. Alia (2004) provides an example of the importance of considering placement in her observation that when major

U.K. daily *The Guardian* provided a rare and important critique of how media cover women in sports, they placed the story in the daily magazine. The significance of such placement, Alia contends, is that "such criticism would have far greater impact it if appeared in the sports pages" (p. 134).

Fourth, by focusing on the "unlawful" acts of protesters, some of the Canadian papers precluded a framing that would have focused on the human rights aspect or the need for environmental justice by the Tribe. Seen from this perspective, both the *Economic* and *Law and Order* frames are the most powerful of Ibarra and Kitsuse's (1993) types of *counterrhetoric*, for the perceived twin needs for social "order" and economic growth consistently override the Tribe's needs in stories that contained these frames. A special note is in order here about the *Globe and Mail*, because their framing was the most unusual: without considering placement, its articles were quite empathetic and included the Tribe's voice (and the voices of First Nations groups) often. The articles had what I can only call a "soft" feel to them, absent of the strident, critical tone taken by the other two papers. However, their placement reveals an important issue with Canadian coverage, for it seems the only way any daily could consider the issue was through the lens of "can it happen here?"

Along these lines, the majority of the Canadian newspaper articles contained a latent thread of apprehension about the Dakota Access Pipeline protests coming to the "oilpatch" in Alberta. This implicit concern is reflected in the two most frequent frames discernible in the papers' coverage: *Economic* and *Law and Order*, for the framing suggests that Indigenous pipeline resistance is either bad for business, a danger to the social order or (often) both. In terms of the *National Post* specifically, the majority of articles were framed by economic concern about what the protests would mean for Canada's ability to sell its oil by transporting it out of the country. Thus, the paper's stories, like those from the other two papers, were infused with an apprehensive tone regarding whether the U.S. protests would come north.

Finally, analysis reveals that, when it comes to the coverage of Standing Rock and the pipeline, when a paper employed an *Economic* frame it often used the *Law and Order* frame as well – a finding that becomes the focal point of the next chapter. It is important to note that an *Economic* framing was not always paired with the *Law and Order* frame, especially for the *Globe and Mail*, whose coverage was different in this regard from the other two papers. This was the only paper included in this analysis – in either the U.S. or Canada – to not employ a *Law and Order* framing to its coverage. In fact, the paper actively avoided it. An article published at the height of the Standing Rock struggle (titled "Pipeline Companies Brace for Clashes; Kinder Morgan in 'Deep' Talks with Police") noted that the corporation had been watching the Standing Rock protests as a way to instruct pipeline construction contractors "should they encounter protesters, drawing lessons from clashes in North Dakota over the $3.7-billion (U.S.) Dakota Access pipeline under construction." The story quotes Chris Bloomer, CEO of the Canadian Energy Pipeline Association (CEPA), as saying that "demonstrators have the right to voice opinions about energy development, but tampering with pipelines puts nearby communities and the environment in danger."

The article, however, then notes that the activists had called the pipeline companies a few minutes ahead of time to warn them of the action they would take. The journalist had called the pipeline corporation, which "confirmed it received the call." In following up with the story and checking facts, the *Globe and Mail* avoided portraying the activists as lawbreakers and told the truth of the story, which was confirmed by none other than the pipeline company itself.

Why did the paper avoid a *Law and Order* framing seen in other papers like the *Calgary Herald* and *Bismarck Tribune*? One reason may be its politics: as noted earlier, the paper typically takes a less conservative tone than others. Another reason may be that the paper has less at stake financially than regional papers. Thus, while the *Globe and Mail* demonstrated an unwillingness to see the Standing Rock struggle as a standalone issue worthy of exploration outside of the economic benefit or detriment to Canada, it contained no direct *counterrhetorics* to the Tribe's concerns as observed in other Canadian papers. Put another way, the paper never negated the sovereignty or the power of the Tribes to dispute a pipeline crossing "traditional territory." Instead, the paper believed that pipeline corporations should work closely with Tribes when it comes to oil exploration and transportation.

As a result of these patterns in coverage, one of the most important findings for the analysis of Canadian news coverage of Standing Rock can be defined as a paucity, where the "paucity" to which I'm referring can be thought of in two ways: the small number of articles dedicated to the pipeline demonstrations, yes, but also in terms of the lack of focus on Standing Rock as a standalone issue worthy of consideration outside the lens of endangerment to social order, oil industry profits, and overall economic growth. Regarding the small number of articles, the question that still needs to be addressed is just *why* the vast majority of Canadian newspapers (not simply the ones included for analysis in this chapter) decided not to cover the events. The answer as to why the vast majority of Canadian newspapers didn't cover the issue in much depth relates to and returns us to the concept of *deep media*.

Media blackouts, media "draws," and *deep media*

Given the worldwide attention the Standing Rock Sioux Tribe and other Indigenous groups received during their resistance to the Dakota Access Pipeline, the very few number of stories from the U.S.'s contiguous northern neighbor seems at first perplexing. The overall lack of coverage is especially intriguing given commercial journalism's consistent focus on acrimonious conflict, which coheres with the "if it bleeds it leads" professional routine. In fact, the protests, which were marked by striking visuals of Native Americans being led away by police, soaked with water cannons in freezing temperatures, attacked by security company dogs, and standing in the mists of tear gas, fulfilled most requirements for commercial news coverage to become interested and engaged. More specifically, the Standing Rock protests would seem to be an alluring topic for mainstream news media in that they first represented a discrete event. Numerous scholars (including Schudson, 2011; Boykoff & Boykoff, 2007; Beder, 1998;

Anderson, 1997; and Hall et al., 1978) recognize the media's tendency to focus on events – or what Schudson (2011) refers to as an "event-centered" distortion. While on a superficial level this may seem like a natural focus, if one considers the importance of covering more complex processes (like climate change or Congress[7]), the nature of the journalistic distortion becomes clear. Snow and Benford (1992) astutely observe that "specific movements within any historical era are tributaries of a more general stream of agitation"[8] and thus episodic coverage without context is a distortion and flaw in and of itself.

In line with episodic reporting is a focus on conflict: the events at Standing Rock contained dramatic visuals that were infused with conflict. Bartholomé et al. (2015, p. 451) note that "Journalists do not merely disseminate conflict frames put forward by political actors, but actively shape when and how conflict appears in the news. Subtle methods of journalistic news production are applied to facilitate, emphasize, and sometimes even exaggerate conflict." Conflict, of course, is not only categorized as "negative" news (another of Schudson's distortions and a draw for commercial media) but also lends itself to rather polarizing "good" versus "bad" guys as recognized by Beder (1998) and Anderson (1997) (of course, which side is which depends on the outlet's perspective). Speaking directly to news coverage of Indigenous groups, Larson (2006) argues that "National news is attracted more to stories involving Native Americans that are sensational, vivid, and conflictual" (p. 109).

Finally, the lack of coverage is intriguing due to the attention the protests received from politicians, from low-level mayors to governors to several U.S. presidents. Canadian politicians also discussed the Dakota Access Pipeline protests. The reason this is "intriguing" is due to what Boykoff and Boykoff (2007) refer to as "authority-order bias," where news stories commonly originate in and revolve around what "official sources" say and do in relation to an event, person, or issue. Official sources tend to predominate the news, and one of the hallmarks of professional journalism is the over-reliance on these types of sources (Schudson, 2011; McChesney, 2008). Given the high level of attention paid to this by a wide variety of official sources (and including the fact that many of the papers cited President Trump and his actions at length), the lack of coverage stands out even more.

This discussion of the results of analysis is where the previous discussions of the patterns and routines of professional journalism (Chapter Two) as well as news media's coverage of environmental issues (Chapter Three) come full circle, for the events at Standing Rock contained many of the well-established "draws" that typically would attract coverage from mainstream news outlets. As Hall et al. (1978) astutely recognize, the news media tend to publish stories that demonstrate what they term "extraordinariness" – demonstrated by a focus on dramatic events that have a negative aspect (usually focused on conflict), that are easily personalized, and that involve powerful players or entities (p. 53). This focus on "extraordinariness" by the news media thus invites the question of why there was almost what could be considered a "radio silence" on the issue, and when it was covered, why it was never considered a standalone issue worthy of exploration and understanding through any other lens than financial interest.

Here I argue that, despite the elements of the Standing Rock story that would normally have provided a media draw, the deep structural flaws in the media system – including those that privilege considerations of profit and are undergirded by corporate influence – served to limit the frequency of coverage as well as skew the tenor of that coverage predominantly towards economic concern. Seen from this perspective, the coverage by Canadian newspapers seems to be influenced by the desire for Canadian economic growth through selling fossil fuels from the Alberta "oilpatch." Part of selling the oil is, of course, ensuring transportation of the product to other markets through pipelines: as a result, the overarching fear expressed by the newspapers seemed to be in regard to whether the protests at Standing Rock would influence environmental groups and First Nations groups in Canada to follow suit. The silence by the vast majority of the newspapers in Canada appears to revolve around a "what problem?" stance through omission, where the hope might have been that lack of coverage of the issue might translate into lack of action up north.

Comparing the Canadian papers with the Bismarck Tribune

Returning to the U.S. study for comparison, it is clear that the *Bismarck Tribune*, located as it is within another large oil-producing region, shared many of the same economic concerns as the Canadian papers. But while the Canadian press may have been able to largely ignore the events to the south, the *Bismarck Tribune* – located at the heart of the conflict – could not. My findings of Canadian newspapers in particular support earlier scholarly research on news media coverage of Indigenous environmental justice issues. As Deacon et al. (2015) have found, Canadian dailies often privilege the idea of economic growth over environmental justice concerns by emphasizing the risks of not protecting and nurturing financial prospects. In addition, they note that environmental justice issues are often "muted" by Canadian daily newspapers, which "overtly ignore and subtly downplay through discourse environmental threats to marginalized local communities faced with pollution" (p. 419). As a result, these stories tend to frame environmental risk to communities of color as "unproblematic – rather than unjust" (p. 426). This is similar to what Haluza-Delay (2007) has observed, namely that: "the environmental justice frame has not caught on in Canada" (p. 559). Similarly, in his analysis of the *Vancouver Sun*, Stoddart (2007, p. 665) contends that the newspaper simplified environmental policy debate into a conflict between the government, the forest industry, and environmental groups, marginalizing the perspective and voices of First Nations groups. From this perspective, it becomes clear that the twin frames of *Law and Order* and *Economic* worked in concert to minimize and reject the concerns of the Standing Rock Sioux Tribe as invalid.

Conclusion

Boykoff (2006, p. 203) notes the important historical role played by the mass media in suppressing dissent in the United States, as they tend to look more favorably on dissident citizens who operate within the system and to disparage dissidents whose oppositional activities challenge sanctioned modes of action. While dissidents are sometimes able to frame issues and grievances in a manner satisfactory to them, they are more often frustrated by what they deem inadequate—and sometimes even derisive—mass-media coverage. Coverage frequently fails to focus on the issues and ideas of social movements and actually deprecates the participants, thereby undermining social movement efforts.

The deep structural problems with commercial media help to explain the news media's suppression of dissent and resistance that isn't seen to be "sanctioned," especially resistance that involves the fight against large interests like the fossil fuel industry.

When it comes to the cultural politics of representation in the news media, Merskin (2014) observes that: "the responsibility for ethical representations rests on the shoulders of those doing the presenting. Lacking clear sociological grounding as truth or falsity, images convey symbolic information that embodies shared cultural understandings" (p. 197). Although distinct in nature, the #NODAPL movement can be situated within a broader movement involving twin concerns: extracting and using fossil fuels at a time when the effects of climate change are increasingly felt, and the rising power and recognition of Indigenous rights. This chapter ends with a quote from Mark Anthony Rolo, who recognizes the power that the news media wield in shaping public perception on issues. Here, the lack of adequate coverage serves to stall the push towards recognizing the needs of Indigenous groups:

> Can American Indians continue to wait for the nation's newsrooms to hire and promote more American Indians to help ensure fair, accurate, and timely news coverage of our community? Obviously the answer is a resounding no. These times are extremely critical for American Indian nations. As tribes push towards more self determination [and] act on land and water rights agreed upon by treaties signed with the federal government, there are those in the private, corporate, and political sector who are doing all they can to eliminate those rights. The stakes are high for Indian people....
>
> *(2000, pp. 9–10)*

Notes

1 After excluding the articles originally written by the Associated Press and Canadian Press, there remained only five articles written by the *Chronicle Herald*, which likely is due to a lengthy union strike by journalists at the paper in 2016 and 2017.

2 The *Calgary Sun* first was published under the name *The Calgary Herald, Mining and Rancher Advocate and General Advertiser.*
3 Retrieved from https://nmc-mic.ca/about-newspapers/circulation/daily-newspapers/
4 The Canadian Encyclopedia. (2010). *Calgary Herald.* Retrieved January 19, 2018, from http://www.thecanadianencyclopedia.com/en/article/calgary-herald/
5 The frame was originally defined in Moore and Tucker (2017).
6 Available at https://nmc-mic.ca/wp-content/uploads/2015/02/Snapshot-Fact-Sheet-2016-for-Daily-Newspapers-3.pdf
7 Schudson (2011) provides the example of former-GOP Congressman Bill Frenzel, who made the observation that Congress is more of a process than a series of punctuated events.
8 For this information, Snow and Benford (1992) cite Turner and Killian's *Collective Behavior* (1987).

References

Alia, V. (2004). *Media ethics and social change.* New York: Routledge.

Anderson, A. (1997). *Media, culture and the environment.* London: Routledge.

Bartholomé, G., Lecheler, S., & de Vreese, C. (2015). Manufacturing conflict? How journalists intervene in the conflict frame building process. *International Journal of Press/Politics, 20*(4), 438–457. 10.1177/1940161215595514

Beder, S. (1998). *Global spin: The corporate assault on environmentalism.* White River Junction, VT: Chelsea Green Pub.

Boykoff, J. (2006). Framing dissent: Mass-media coverage of the global justice movement. *New Political Science, 28*(2), 201–228. 10.1080/07393140600679967.

Boykoff, M. T., & Boykoff, J. M. (2007). Climate change and journalistic norms: A case-study of US mass-media coverage. *Geoforum, 38*(6), 1190–1204. 10.1016/j.geoforum.2007.01.008.

Deacon, L., Baxter, J., & Buzzelli, M. (2015). Environmental justice: An exploratory snapshot through the lens of Canada's mainstream news media. *Canadian Geographer / Le Géographe Canadien, 59*(4), 419–432. 10.1111/cag.12223.

Entman, R. M. (1993). Framing: Toward clarification of a fractured paradigm. *Journal of Communication, 43*(4), 51–58. 10.1111/j.1460-2466.1993.tb01304.x.

Gamson, W., & Modigliani, A. (1987). The changing culture of affirmative action. *Research in Political Sociology. A Research Annual 3,* 137–177. ISSN 0895-9935.

Gamson, W., & Modigliani, A. (1989). Media discourse and public opinion on nuclear power: A constructionist approach. *American Journal of Sociology, 95*(1), 1–37.

Hall, S., Critcher, C., Jefferson, T., Clarke, J., & Roberts, B. (1978). *Policing the crisis: Mugging, the state, and law and order.* New York: Holmes & Meier.

Haluza-Delay, R. (2007). Environmental justice in Canada. *Local Environment, 12*(6), 557–564. 10.1080/13549830701657323.

Ibarra, P., & Kitsuse, J. (1993). Vernacular constituents of moral discourse: An interactionist proposal for the study of social problems. In G. Miller & J. Holstein (Eds.), *Constructionist controversies* (pp. 21–54). New York: Walter de Gruyter.

Larson, S. G. (2006). *Media & minorities: The politics of race in news and entertainment.* Lanham, MD: Rowman & Littlefield.

McChesney, R. W. (2008). *The political economy of media: Enduring issues, emerging dilemmas.* New York: Monthly Review Press.

Merskin, D. (2014). How many more Indians? an argument for a representational ethics of Native Americans. *Journal of Communication Inquiry, 38*(3), 184–203. 10.1177/0196859914537304.

Potter, J. (2014). The National Post. *The Canadian Encyclopedia.* Retrieved from https://www.thecanadianencyclopedia.ca/en/article/the-national-post/

Potter, J., Doyle, R., & Yusufali, S. (2015). Globe and Mail. *The Canadian Encyclopedia.* Retrieved from https://www.thecanadianencyclopedia.ca/en/article/globe-and-mail/

Rolo, M. A. (2000). Introduction. In *The American Indian and the media* (2nd ed., pp. 8–13). St Paul, Minnesota.

Schudson, M. (2011). *The sociology of news* (2nd ed.). New York: W.W. Norton & Company.

Snow, D., & Benford, R. (1992). Master frames and cycles of protest. In A. Morris and C. M. Mueller (Eds.), *Frontiers in Social Movement Theory* (pp. 133–155). New Haven, CT: Yale University Press.

Stoddart, M. C. (2007). 'British Columbia is open for business': Environmental justice and working forest news in the Vancouver Sun. *Local Environment, 12*(6), 663–674. 10.1080/13549830701664113.

6

LAW AND ORDER

From civil rights to Nixon to Trump, a trope in revival

As I began to write this chapter in mid-February 2018, U.S. Attorney General Jeff Sessions made the following statement to a crowd of law enforcement personnel gathered in Washington, D.C.:

> I want to thank every sheriff in America. Since our founding, an independently elected sheriff has been the people's protector, who keeps law enforcement close to – and accountable to – people through the elective process. The office of the sheriff is a critical part of... of the *Anglo-American heritage* of law enforcement. We must never erode this historic office. I know this, you know this.
>
> *(Vazquez, 2018, emphasis added)*

Sessions' off-the-cuff remarks on law enforcement's "Anglo-American heritage" (not included in the earlier publication of his official notes by the U.S. Department of Justice[1]) are intriguing because they provided a direct window into a certain way of thinking about the history and purpose of law enforcement in America: it is there to secure social order through the rule of law, and its history is marked by Anglo-American governance and interests. The comments made by Sessions are particularly germane to this chapter due to the persistent questions that have been raised about the impact of "law and order" approaches on many communities of color. As such, they are particularly relevant to the current research on how the media chose to cover Standing Rock.

Following on the heels of the framing analysis of newspapers in the U.S. (Chapter Four) and Canada (Chapter Five), this chapter continues the thread by outlining the history and cultural politics of "law and order" rhetoric in the U.S. as it relates to news coverage of the events at Standing Rock. The results of the framing analysis revealed that *proximity* mattered – meaning that the newspapers closest to areas involved with fracking of oil and gas were more likely to cover the events surrounding the pipeline through two primary frames: *Economic* and *Law and Order*. In fact, when newspapers tended to have a high number of stories with *Law and Order* frames, they tended to also contain many articles with *Economic* frames. Perhaps this is unsurprising, given that the political and social rhetoric of "law and order" is often seen to be intertwined with economic concerns (Flamm, 2005; Friedman, 2000; Hall et al., 1978; Platt, 1994). On the obverse side of this coin, newspapers that tended to avoid *Economic* framing most often created articles that exhibited *Empathy* or *Environmental Justice* frames that underscored the environmental and cultural needs of the Standing Rock Sioux Tribe. That the *Economic* framing appears to be intimately connected to *Law and Order* is explored later in this chapter as it relates to racial identity and cultural politics.

The next step in this research, which is the focus of this chapter, is to explore the background for this media framing. In other words, the intent here is to contextualize the heavy emphasis by some newspapers on the need for social order, increased policing, and economic growth as it relates to both history and contemporary electoral and cultural politics. This chapter begins with an exploration of the rhetoric of "law and order" in U.S. history, progresses to a discussion of where we find ourselves now as a nation with this approach, and then addresses the significance of a law and order perspective on the Standing Rock movement. The chapter ends with a discussion of the relationship between *deep media*, environmental issues, and the politics of racial representation in the U.S.

What is law and order?

This section outlines the origins and scope of "law and order" rhetoric in the U.S. as a way to identify the underlying roots of – and connections to – some of the coverage of the #NODAPL movement. The intent here is not to reproduce the complex, critical scholarship on mass imprisonment often referred to as "carceral research" (see, for instance, Pickett & Ryan, 2017; Gottschalk, 2015, 2006; Gross, 2015; Gandy, 2014; and Alexander, 2010), but instead to build on scholarship that recognizes racial and economic components in the "law and order" posturing visible in U.S. politics and mainstream media. This line of inquiry aims to provide not only a "how" but also a potential "why" newspapers covered #NODAPL the way they did, while also exploring the deeper consequences of *Law and Order* and *Economic* framing of environmental justice movements.

The history of law and order rhetoric in the U.S.

Some of the first uses of the phrase "law and order" in the U.S. stretch as far back as the writings of the earliest presidents. In *The Works of John Adams, Second*

President of the United States, Adams stressed the need, in the early days of the formation of the nation, for law and order:

> great was the disappointment to discover... that independence had not done all that was hoped of it, that the people were not prosperous, that law and order were not so well established as they had been in the colonial days, that instead of improving with time, things were visibly growing worse.
>
> *(1856, pp. 439–440)*

Adams also stressed the need for law and order in his *Proclamation 8* in 1799 during a period of resistance to taxes in Pennsylvania, when he demanded that the "insurgents" should disperse, warned others against helping them, and claimed that he would call upon "military force" to enable the laws to be "duly executed" (Minot & Sanger, 1863, p. 757). In his *Memoir, Correspondence, and Miscellanies*, Thomas Jefferson (1829) praised men who were guided by reason in choosing the side of "law and order" (pp. 369–370). Jefferson also contended that this choice enjoyed near unanimity, noting that: "everyone...is interested in the support of law and order" (p. 230).

Centuries later, the phrase experienced a significant revival. In his book "Playing with Fire," Lawrence O'Donnell (2017, p. 208) quotes the well-known speech "A Time for Choosing" given by Ronald Reagan[2] as he stumped for the Republican presidential candidate Barry Goldwater in 1964:

> You and I are told we must choose between a left or right, but I suggest there is no such thing as a left or right. There is only an up or down. Up to man's age-old dream – the maximum of individual freedom consistent with order – or down to the ant heap of totalitarianism.

O'Donnell observes that the particular phrase *consistent with order* represented

> the first draft of what was to become "law and order" in 1968. Republicans knew what "order" meant – no more riots, no more demonstrations and protests... the stuff that was tearing the country apart in the eyes of conservatives. For them, order had completely broken down, and it had to be restored with no compromise.
>
> *(p. 208)*

Goldwater himself embraced a law and order stance, making it a key component of his campaign against President Lyndon B. Johnson. In his compelling and meticulously researched book about the rise and fall of law and order political rhetoric in the 1960s, Flamm (2005) pinpoints Goldwater's speech at the Republican National Convention as the moment that "law and order became an important part of national political discourse" (p. 31).

Underpinning the growing political focus on law and order, Hall et al. (1978) observe, were three interrelated beliefs: that there were "massive" increases in the crime rate, that criminals were "getting off lightly," and that the only tactic in response was "to get 'tough'" (p. 9).[3] Similarly, Flamm (2005) attributes the rise in popularity at this time in history to a few different factors, including "street crime, urban riots, and political demonstrations" (p. 2). First, he points to the "Long, Hot Summer" of 1967 as a time of numerous riots in over 100 U.S. cities that grew out of various racial tensions. He highlights two riots in particular, noting that the police raid of a "blind pig" (unlicensed bar) in Detroit[4] and the arrest and beating of a black taxi driver in Newark caused racial tensions over police treatment to boil over into looting, violence, and standoffs with the police. Reaction to the riots, Flamm observes, reveals a particularly important racial element – namely that while many black Americans attributed the unrest during the "Long, Hot Summer" to systematic discrimination, police brutality, unemployment, and inadequate housing, white Americans believed the riots to be planned and "organized," and thus rejected the idea of police brutality playing a significant role. Second, Flamm notes that the murder rate in the U.S. almost doubled between 1963 and 1968, which played into the fears of "anxious whites" regarding their sense of personal safety (p. 2). Finally, the mass protests over the war in Vietnam precipitated the more widespread societal adoption of a law and order stance. But, even considering these factors, there was more. The slowing economy of the 1960s precipitated frustration among white workers "and made them more receptive to messages that blamed others – especially minorities – for their predicament" (Flamm, 2005, p. 7). The competition posed by people of color in the workplace created a palpable sense of insecurity that created "fertile soil," as Flamm puts it (p. 5), for the law and order rhetoric that President Nixon embraced later in the decade.[5]

On October 4, 1968, as Richard Nixon ran in a contentious and close race with Democrat Hubert Humphrey, *Time* magazine published an issue titled "Law and Order." The image on the cover depicted a stylized drawing of a policeman wearing a blue hat bedecked with three law enforcement badges: "Metropolitan Police," "Constable," and "Sheriff." Halfway down the image (blending with the lower half of the policeman's face) was a series of raised fists, some carrying signs or sticks, with one pointing a gun at the policeman's head. The article begins by noting that "The presidential campaign of 1968 is dominated by a pervasive and obsessive issue. Its label is law and order." As Graham (2016) has observed,

> "Law and order" was a potent mantra for Nixon during the 1968 campaign. The nation seemed like it was on fire – and parts of it actually were. In April, Martin Luther King Jr. was assassinated in Memphis. In June, Robert F. Kennedy was killed. Riots swept through major cities. After an unprecedented period of low crime after World War II, the violent-crime

rate was in the midst of a sharp rise that would taper off slightly in the early 1980s before rising again, finally peaking in 1991. Americans were afraid, and Nixon's promise to get tough on crime and stop the growing violence exercised a powerful appeal.

In a 1968 campaign speech titled "Order and Justice under Law," Nixon elucidated what was to be his law and order stance as president, which involved condemnation of violent protests, a denial that poverty was a cause for civil unrest, another denial "that law and order was a racial slogan," and accusations that the U.S. Supreme Court was privileging defendants' rights over that of law enforcement (Flamm, 2005, p. 173). Nixon's law and order stance flourished in the heady mixture of Vietnam War protests, the 1967 riots, the civil rights movement, the "free love" movement, and increased drug use among young Americans. As Blumenthal (2016, p. 29) observes, "law and order had become a powerful weapon for conservatives in the 1960s who blamed liberal leaders for the rise in street crime, drug use and campus unrest."

Although he didn't particularly distinguish between them, Nixon, adopted a law and order approach to his rather unsuccessful[6] war on drugs, street crime, protests, and social unrest over the Vietnam War (Blumenthal, 2016). A recent article in the *Canadian Press* summarizes the context for Nixon's law and order approach concisely:

> Nixon capitalized on chaos at the Democratic convention, where anti-Vietnam war protesters were clubbed and tear gassed by police. Those events came just after African-American neighbourhoods had been torched in different cities, following the assassination of Martin Luther King. Nixon responded by staging a campaign rally at the site of the convention – Chicago – where he promised change: "We're going to have law and order." His TV ads declared: "We have to rebuild respect for law in this country."
>
> *(Panetta, 2016)*

Nixon's reaction to these various events, which included the "command and control" strategy that was the hallmark of the law and order approach, is telling in its refusal to acknowledge the context in which all these events took place. There is no admission, in his response, of the sadness and anger surrounding the numerous deaths of U.S. service members during the controversial war or the sense of frustration and outrage in African-American communities over the loss of numerous political leaders who were driving the civil rights movement. There is no recognition of the deep stressors of poverty or unemployment, of police brutality, or the need for civil rights in Nixon's response – only the need to control the unrest. Because of this, the political rhetoric of law and order, especially as manifested in the 1960s onward, consistently has been criticized as having distinctly racial undertones.

Law and order, racial identity, and the civil rights movement

In Michelle Alexander's well-known and oft-cited book *The New Jim Crow* (2010), she observes that

> at the same time that civil rights were being identified as a threat to law and order, the FBI was reporting fairly dramatic increases in the national crime rate…to make matters worse, riots erupted in the summer of 1964 in Harlem and Rochester, followed by a series of uprisings that swept the nation following the assassination of Martin Luther King, Jr. The racial imagery associated with the riots gave fuel to the argument that the civil rights for blacks led to rampant crime.
>
> *(pp. 41–42)*

A 1968 article in the *New York Times* ("Negro Leaders See Bias in Call of Nixon for 'Law and Order'") described the reaction by many people of color to the "law and order" approach during this time. The story cites an NAACP board member's contention that "Nixon's appeal for law and order has special meaning when he uses it. I'm all for law and order, but he is trying to get the support of the white backlash people around the country." The phrase, the journalist noted, evokes a sense of "unequal justice" and racial inequality (Johnson, 1968). Due to the focus on ensuring white citizens' personal safety (as well as a sense of societal order), even early calls for law and order were deemed racist, for they ignored "root causes like poverty and unemployment" (Flamm, 2005, p. 2). In addition, Nixon didn't help assuage fears that law and order *wasn't* about race: as Cohen (2016, p. 311) notes, Nixon referred to the lawlessness of the city and its threat to the white suburbs:

> "Without progress," he instead darkly warned, "then the city jungle will cease to be a metaphor…It will become a barbaric reality and the brutal society that now flourishes in the core cities…will annex the affluent suburbs." These images were racially charged and directly pitched to white audiences.

Adding to this racially charged verbal rhetoric were visual cues: a *National Public Radio* (Nunberg, 2016) article noted that some of Nixon's most popular campaign ads depicted a white woman fearfully walking down a dark city street, "where the assailants who might be lurking in the shadows presumably weren't hippies or demonstrators."

Ideas about the need for strict social order were not limited to the arena of electoral politics. Politicians' refusal to acknowledge underlying causes for social unrest combined with conspicuous calls for "order" also were seen clearly in the response by the white clergy in the American south to the Birmingham Campaign in the spring of 1963. The campaign in Alabama involved various peaceful demonstrations, including restaurant sit ins, boycotts, and marches, but was met with police force that included police dogs and high-pressure fire hoses used on protesters – creating some of the most memorable and iconic images from the civil rights movement. Many white southern Christian leaders had written to Martin Luther King, Jr. that he needed to be patient – that justice would come eventually but that the time wasn't right for these protests, that King and his followers should be patient, and that order should be upheld. In his now famous *Letters from Birmingham Jail*, King penned a response to them where he questions why his "fellow clergymen" would choose to privilege the idea of social *order* over the need for social *justice*:

> I had hoped that the white moderate would understand that law and order exist for the purpose of establishing justice and that when they fail in this purpose they become the dangerously structured dams that block the flow of social progress.
>
> *(2000, p. 97)*

In another passage, King draws attention to the need to understand the underlying cause for the protests, observing that

> you deplore the demonstrations taking place in Birmingham. But your statement…fails to express a similar concern for the conditions that brought about the demonstrations….It is unfortunate that demonstrations are taking place in Birmingham, but it is even more unfortunate that the city's white power structure left the Negro community with no alternative.
>
> *(2000, p. 87)*

King's rejection of the white Southerners' call for "order" and his push instead for justice is mirrored in *The New Jim Crow* (2010). Alexander notes that from the 1950s through the 1960s,

> conservatives systematically and strategically linked opposition to civil rights legislation to calls for law and order, arguing that Martin Luther King Jr's philosophy of civil disobedience was a leading cause of crime. Civil rights protests were frequently depicted as criminal rather than political in nature… [and were linked] to the spread of crime.
>
> *(p. 41)*

Alexander perceives that law and order rhetoric and practice is an integral part of a broader system of structural racism marked by "racial stigmatization and permanent marginalization" (p. 12). Specifically, she notes that the rhetoric, which both marginalized and stigmatized people of color,

> was first mobilized in the late 1950s as Southern governors and law enforcement officials attempted to generate and mobilize white

opposition to the Civil Rights Movement. In the years following Brown vs the Board of Education, civil rights activists used direct-action tactics in an effort to force reluctant Southern states to desegregate public facilities. Southern governors and law enforcement officials often characterized these tactics as criminal and argued that the rise of the Civil Rights Movement was indicative of a breakdown of law and order. Support of civil rights legislation was derided by Southern conservatives as merely 'rewarding lawbreakers'.

(p. 40)

The recognition that the civil rights movement was perceived as a breakdown of social order aligns easily with the stance taken by John Adams (on tax protesters), Reagan (street crime), and Nixon (protests, riots, drug use, and street crime all rolled together) in the sense that the solution proposed by all of these politicians involves the use of *force* to restore order: not compromise, not negotiation, but force. Thus, the history of law and order as used in the U.S. is problematic on myriad levels, not the least of which is the racial element in the rhetoric and practice of it. And although the focus thus far in this chapter has been on the history of the law and order stance, U.S. President Trump has revived the Nixonian rhetoric as a key part of his platform.

Where we are now: racial identity, justice movements, and Trump as the new law and order president

As is now well known, in the U.S. 2016 presidential election, Donald Trump and his running mate Mike Pence were swept into the most powerful political positions in the country. Trump's law and order approach to perceived problems of crime and unrest – combined with his "I alone can fix it" mantra – constituted a large part of his electoral platform. As has been well chronicled by now, Trump specifically became enamored with former president Nixon's law and order approach to dealing with various social issues, which was especially evident on the night of his nomination at the Republican National Convention:

> When Donald Trump strides to the stage Thursday night to accept the presidential nomination of Abraham Lincoln's political party, his address will borrow heavily from a more recent speech by another party nominee: Richard Nixon – 1968. The presumptive Republican nominee's campaign manager said Monday that Trump studied past speeches while preparing his own, and was most captivated by the one Nixon delivered in a year with similar political overtones to this one. Back then the president-to-be promised law and order in a speech that referenced the country's dual crises: race-related riots at home that left inner cities torched, and the bloody conflict abroad in Vietnam…. Trump has embraced the law-and-order theme.
>
> *(Panetta, 2016)*

As Kranish and Fisher (2016) describe in their biography *Trump Revealed*,

> Trump's... program presented TV viewers with a vision of a country in deep trouble, unsafe, weak, governed by rigged systems and by people who had dishonest, even evil intentions. Speakers described a country...nearly at the mercy of what Trump called 'barbarian' terrorists. It was a gloomy picture of a nation in decline.
>
> *(2016, p. 339)*

In a speech earlier in 2016, Trump claimed that "We must maintain law and order at the highest level or we will cease to have a country, 100 percent.... I am the law and order candidate" (Nelson, 2016). During one of the U.S. presidential debates at Hofstra University, Trump again claimed that he would be the "law and order president" while adding that he advocated for the highly controversial "stop and frisk" police tactic. In March 2018 during one of his rallies, Trump called for the death penalty for drug dealers, revealing that his version of law and order is even more extreme than Nixon's.

As it was back in 1968, many perceive a racial component to Trump's new version of law and order. The *Philadelphia Tribune* held a viewing of Trump's debate with Hillary Clinton with members of the community. One Philadelphia citizen contended that "I think [the phrase] 'law and order' is a code-word for keeping Negroes in line.... 'Law and order' is clearly a message to elements of America who are fearful that there is a problem" (Lee, 2016). One element to a law and order perspective is not only a conceptualization of the threat that criminality imposes on society as a whole but also an explicit support for law enforcement. In a video his campaign posted to Facebook in July of 2016 (when he was still a presidential hopeful on the campaign trail), Trump addressed the death and injury to police officers in Dallas, Texas, noting that "we must stand in solidarity with law enforcement, which we must remember is the force between civilization and total chaos. Every American has the right to live in safety and peace.... We will make America safe again."

When he became president, some of Trump's earliest executive orders provided official support for law enforcement and highlighted the need for law and order. Two in particular stand out, including the *Presidential Executive Order on Preventing Violence Against Federal State, Tribal and Local Law Enforcement Officers* (which calls for stiff penalties and sentences for those who harm law enforcement personnel) and the *Executive Order on a Task Force on Crime Reduction and Public Safety* (which directs the U.S. Attorney General to create a task force to address "illegal immigration, drug trafficking, and violent crime").[7] As Harden (2017) accurately points out, these orders (among others) put an increased burden on people of color as well as immigrants to the U.S. while simultaneously exhibiting support for law enforcement and increased policing. Regarding the first, it is widely recognized that there already are strong penalties in place for injuring police, so Trump's executive order simply underscores the need to punish those accused of a crime; regarding the second, a crackdown on "illegal immigration" does not help to reduce crime rates, since undocumented immigrants are not responsible for the majority of crimes (and instead are much more likely to be victims of crime). As Bernat (2017) observes, "quantitative research has consistently shown that being foreign born is negatively associated with crime overall and is not significantly associated with committing either violent or property crime" (p. 1). While Trump's executive orders didn't create specific policies in and of themselves, they served as a highly visible public affirmation of Trump's support for law enforcement and the need for strict policing to maintain social order.

Trump's concerns regarding the need for "law and order" exist within his broader populist appeal – an approach that has been seen in several other countries in the last few years, including France (with the National Front party leader Marion Le Pen) and the United Kingdom (with its "Brexit" vote). Although it certainly is worthwhile to explore the underlying reasons for the global rise of this type of "authoritarian populism," which can include a "cultural backlash" from fears relating to economic security and increasing rates of immigration (Inglehart & Norris, 2017),[8] much of that discussion would take us outside the scope of this book and away from an elucidation of the complexities behind the revival of the more specific rhetoric of law and order that exists not only within Trump's populist approach but also in the approach taken by many mainstream media outlets to the Standing Rock protests.

Trump, Native Americans, and the Dakota Access Pipeline

In early 2017, a few weeks into his term as president, Trump signed executive orders intended to advance the completion of both the Dakota Access and the Keystone XL Pipelines. In *Memorandum for the Secretary of the Army*, Trump directs the U.S. Army Corps of Engineers to approve of the continuation of the Dakota Access Pipeline construction. This action by Trump did not surprise many environmental organizations of Native American groups. For environmental groups, Trump's intentions were clear when he appointed Scott Pruitt, who had fought many of the Environmental Protection Agency's regulations and often accused the government organization of having an activist agenda.[9] Native American tribes as well recognized a long history with then-businessman and casino owner Trump.

The idea that there may be a connection between Trump's stance on Native American rights and his law and order approach is best described by Boburg (2016), who writes that

> The racially tinged remarks and broad-brush characterizations that Trump employed against Indian tribes for over a decade provided an early glimpse of the kind of incendiary language that he would use about racial and ethnic groups in the 2016 presidential campaign.

Trump's contentious history with Native American groups was evident in earnest in the 1990s when many Indigenous nations in the U.S. (especially on the east coast) were starting casinos. Trump seemed to believe that the Native American-run casinos were his primary competitors, and he chafed at the fact that Native American casinos did not need to pay taxes if the casino was located on reservation land. According to the *Washington Post*, Trump spent

> more than $1 million in ads that portrayed members of a tribe in Upstate New York as cocaine traffickers and career criminals. And he suggested in testimony and in media appearances that dark-skinned Native Americans in Connecticut were faking their ancestry.
>
> *(Boburg, 2016)*

In particular, Trump's comments to the *House Subcommittee on Native American Affairs* indicated that he did not believe those who opened Native American casinos were authentically of Indigenous "blood," stating that "I will tell you right now: they don't look like Indians to me." Trump then opined to the subcommittee that they were privileging Native American rights over his, which he found to be "discrimination."

In April 2018, the news outlet *Politico* published a story noting that the Trump administration was considering categorizing Native American tribes as a separate race rather that a sovereign nation. To do so would have deep ramifications for the ability of Indigenous groups to self-govern as well as retain their access to much-needed health care services. The move, while unwelcome from many Native American groups, was not particularly a surprise due to Trump's history of antagonism with them.

Despite Trump's decades-long tension with Native American groups, and despite the fact that Trump is the most prominent U.S. politician currently touting the need for law and order, he did not single-handedly revive this approach but instead simply tapped into an existing societal sentiment that has waxed and waned in the U.S. for centuries. Trump was not in power during the numerous protests by people of color against police brutality in places like Ferguson, Missouri in 2014. Neither was he president when Heather Mac Donald's controversial book *The War on Cops: How the New Attack on Law and Order Makes Everyone Less Safe* was published in 2016. In the book, Donald makes a case for what she refers to as the "Ferguson Effect," where violent crime is surmised to rise because law enforcement personnel (after the Ferguson protests) became wary of enforcing the law due to fears of increased public scrutiny. The reasoning underpinning the "Ferguson Effect" originated in the backlash to the "Long, Hot Summer" in 1967: the cause of the violent protests is not due to persistent racial discrimination and police brutality (as many black Americans saw it) but instead represents a lack of social order and the need for stricter law enforcement (as many white Americans saw it) (Flamm, 2005). Most relevant to this analysis, Trump was not president when reports began to surface in North Dakota that law enforcement officers were using concussion grenades, water cannons, and tear gas on the Standing Rock protesters. He was not in office when many news outlets chose to cover the #NODAPL resistance through the lens of law and order. Instead, the law and order approach taken by many media outlets, the North Dakota governor's office, and local law enforcement speaks more to the existence of a system in place that privileges a law and order approach to solving conflict than it does to the actions or beliefs of a particular politician. Put another way, while Trump may be leading the most recent dance of law and order in this country, the tune has been in place for centuries, and we need to look both backwards (through history) and forwards (to modern social movements like #NODAPL) to more clearly understand this phenomenon.

Seeing law and order coverage through the lens of moral panics and *deep media*

Vincent Sacco, in *When Crime Waves* (2005), recognizes the media industry's unswerving focus on crime: "that crime and policing are key themes in all forms of media – news and entertainment – seems beyond dispute.... Irrespective of the historical period or the type of medium, crime and crime control provide central forms of narrative" (p. 80). Here, Sacco's point about our "cultural obsession" with crime provides important context for many news outlets' law and order approach to reporting on Standing Rock – it is neither surprising nor is it new. Much of this focus, however, relates to deeper issues relating to disparities in power and influence, especially when it comes to Indigenous protesters. A law and order perspective on reporting crime can be seen to manifest itself from the "top downwards," and is thus directly related to contemporary cultural politics (Hall et al., 1978, p. 218). Noted earlier, Stuart Hall and his colleagues (1978, p. 69) identified three common formats for crime news: police statements regarding particular cases; crime reports (usually with statistics and discernible trends); and stories solely focused on court case proceedings. Considering the media coverage of the Standing Rock protests in Canada and the U.S., it is clear that the *Bismarck Tribune* and the *Calgary Herald* mostly focused on these formats, in that they included many statements from law enforcement about the pipeline resistance, and – especially for the *Bismarck Tribune* in the latter half of its coverage – focused almost exclusively on the court cases involving protesters, often following them for more than a year through the lengthy court processes and official decisions.

Moral panics, law and order, and the news media

Researchers like Hall et al. (1978) and Platt (1994) perceive a type of "moral panic" in media coverage of crime, indicating that as researchers they are more concerned with the *reactions* to crimes like these than they are with the crimes themselves. In other words, by highlighting the subjective nature of news reports of "crime," they recognize that it reveals more about power structures in society and current cultural and political trends that it does about criminals or the rates of crime. Muckraking journalist Lincoln Steffens' provocative 1928 article chronicling how he made a "crime wave" in New York City in the early 20th century speaks to both the subjective nature of news reports as well as their power. He describes the public panic that ensued when he created a particularly sensationalistic story of a street crime (involving policemen unwittingly helping burglars) while writing for the *New York Post* – a story that garnered so much attention that editors at other papers encouraged their journalists to follow suit. What resulted was a journalistic "one-upmanship" by competing papers to craft attention-grabbing reports on crime, which created a public panic by the public until then-Police Commissioner Teddy Roosevelt demanded that he stop:

> Every now and then there occurs the phenomenon called a crime wave. New York has such waves periodically…and they sweep over the public…. Their diagnoses and remedies are always the same: the disease is lawlessness; the cure is more law, more arrests, swifter trials and harsher penalties. The sociologists and other scientists go deeper into the wave; the trouble with them is they do not come up.
>
> *(Steffens, 1928, p. 416)*

Steffens clearly recognizes not only the influence of news reporting on public opinion about crime (whether real or imagined) but also what often is presented as the cure: arrests, punitive measures, and "more law" to restore social order. The creation of moral panics originates within the interlaced and interdependent structures of the government, the police, and the media (Hall et al., 1978). As such, press coverage often complements or can be seen to outright support the efforts of the state to quell crime (or, at least, the perception of crime). With this understanding, then, we move to exploring the interconnection between the media, the state, and social movements in relation to coverage of the Standing Rock protests.

Law and Order framing and the suppression of dissent at Standing Rock

News media coverage of social movements always involves a power struggle, and those who engage in collective action and social movements often find themselves in a tight spot. They need to get media coverage to gain public attention and support for their cause, which means they need to participate in activities that would be deemed "newsworthy"; however, once the media start paying attention, actors in social movements often struggle for the right kind of coverage. The "right" type of coverage is that which gives voice and credence to the movement's concerns – that is, the motivation for taking action in the first place. But getting voices and perspectives recognized by the mainstream media is a formidable challenge.

In *The Suppression of Dissent: How the State and Mass Media Squelch USAmerican Social Movements*, Boykoff (2006) explores how the government and the mainstream press often work in concert to suppress resistance from various political and social actors, observing that "in a democratic society, where open coercion is not an option that can be used too many times without losing legitimacy, the state must come up with subtler ways to maintain social control" (p. 7). In the course of his research, Boykoff systematically identifies the importance of examining the media's role in social change and resistance:

> The mass media constitute a crucial site for the construction of social reality, an ever-unfolding discursive locale that deeply influences public opinion on social issues and significantly delimits societal assumptions and public moods. As such…the mass media play an important role in carving the course of history.
>
> *(p. 193)*

This is important, he reminds us, because of the key historical role that the news media (as well as the government) have played in what he terms "the suppression of dissent" in the U.S.

Boykoff places emphasis on framing as an integral way for the mass media to subtly undercut social movements by observing that the resistive actions of social movements most often are framed in terms of criminality. But in what ways does this suppression occur? According to Boykoff, it often comes from the state as a form of direct violence, employment deprivation, public prosecutions and hearings, surveillance, infiltration, harassment arrests, and "extraordinary rules and laws." The elimination of dissent/social movements also occurs through the actions of the media, which include manipulation, "demonization," "deprecation," and "false balance." Just why this occurs is supplied by Platt (1994), who observes that "the current political preoccupation with crime and justice has little to do with either…. What poses as moral outrage about crime is in fact a recognition of the weakening political authority of the state" (p. 4). Platt's recognition speaks to the need to consider the law and order framing of Standing Rock in relation to the seemingly symbiotic relationship between the government (and its interests) and the role played by commercial media outlets.

Regarding the interaction between the state and the Dakota Access Pipeline company, reports began to surface early in 2017 that the corporation was working directly with law enforcement. As reported in the *Associated Press*, "A private security firm hired by the developer of the $3.8 billion Dakota Access pipeline conducted an aggressive, multifaceted operation against protesters that included a close working relationship with public law enforcement, documents obtained by an online magazine indicate" (Nicholson, 2017). Specifically, criticisms have been leveled at the pipeline corporation positioning a liaison directly in the law enforcement operations center (Nicholson, 2017) as well as court prosecutors and law enforcement using aerial footage from a private security plane operating in a no-fly zone to obtain convictions in felony cases surrounding the protests (Brown et al., 2017).

If stories emerged about close cooperation between law enforcement and the pipeline corporation, none were ever made concerning what might be considered a link between the *Bismarck Tribune*, the pipeline corporation, and the state. Professional news outlets tend to pride themselves on journalistic integrity and independence, and so perhaps this is why no links were ever drawn. However, the paper consistently supported law enforcement efforts to suppress the Standing Rock protests in a few different ways. First, it reduced most claims of law enforcement brutality to a "he said, she said" format that is the hallmark of professional journalism, even though the paper was local and often had journalists on the scene to help tell the truth of the matter. Second, once the pipeline construction began and the oil eventually began to flow through the pipes, the newspaper maintained a clear and consistent focus on the number of arrests of protesters, accompanying these with photos of the downturned faces of protesters next to law enforcement personnel or depicting them in court. Third, even if no arrests were made, the paper often chronicled the "unlawful" or violent behavior of the protesters, highlighting damage to pipeline equipment or violence towards law enforcement. Finally, while the paper did include the Standing Rock Sioux Tribe's perspective consistently, most often the story of the resistance efforts was told through the perspective of law enforcement or the court system through frequent quotes providing information about the criminal acts of the protesters (especially in the final phase of its coverage).

Ultimately, the *Law and Order* frames employed by both the *Bismarck Tribune*, *National Post*, and *Calgary Herald* loosely followed what Hall et al. (1978) term a "signification spiral" in several ways. Coverage through the lens of law and order typically identified specific areas of concern regarding the behavior of the protesters. News stories also tended to identify a "subversive minority" within the protesters (e.g., those whose actions were particularly "unlawful" or violent). Often this "lawless" behavior was linked to other societal problems – most commonly the need for social order and the protection of law enforcement personnel. News articles also contained the idea of an escalating threat that created a "prophesy" of worse to come if the lawlessness was not stopped. Finally, much of the coverage included stern calls for firm steps to end the behavior (Hall et al., 1978, p. 223). Much of this type of coverage by the *Bismarck Tribune* in the last phase of its coverage (2017–2018) did not seem particularly sensationalistic (with detailed accounts of "criminal" acts); instead, the paper published seemingly bland, formulaic reporting on the numerous court cases surrounding the alleged crimes of the protesters. But this type of "just-the-facts" coverage belies a deeper, more troubling lens through which the paper's audience was consistently invited to view the protesters (as potential "criminals"), especially in light of the omission of the Tribe's voice. And, as the framing research revealed, this law and order lens was paired with another lens, which is economic.

Deep Media, Economics, *and* Law and Order

Although the focus of this chapter is on the cultural politics of law and order as they relate to environmental justice issues, the results of the framing analysis presented in Chapters Four and Five revealed that not all news outlets consistently employed a *Law and Order* frame when covering the Standing Rock protests. Instead, other common frames included *Professional Journalism, Economic, and Environmental Justice*. Some patterns, however, did emerge, including that those news outlets that tended to focus on the economic benefits of the pipeline also often employed a law and order perspective, which was especially the case with local dailies from oil-rich regions like the Alberta "oilpatch" (*Calgary Herald*) and the Bakken shale fields (*Bismarck Tribune*). At this juncture, perhaps this is unsurprising, for one of the central points made by this research (and in myriad other political economy studies) is that journalism is fundamentally structured by commercial interests and corporate media ownership. Understanding the economic undergirding to law and order is essential, for both frames served to undercut the environmental justice needs of the Standing Rock Sioux Tribe by displacing the focus from environmental justice to the need for social order and economic growth. This recognition elucidates the ties between government and corporate influence that is at the heart of the conceptualization of *deep media*.

In *Static,* journalist Amy Goodman (2006) critiques the U.S. media as a "megaphone for those in power" (p. 9). Citing, in just one example, news outlets' uncritical coverage of the period leading up to the 2003 war in Iraq, she writes that "the Bush administration called the tune, and the *New York Times* danced" (p. 9). Around this same time, MSNBC famously fired talk show host Phil Donahue, who had one of the top-rated talk shows on the network, as the lead up to the war began. The reasoning provided by the network was that Donahue's criticism of the war created a difficult "public face" for the corporation. The *New York Times* famously critiqued its own coverage of the lead up to the war, finding that it relied too heavily on official sources without properly investigating their claims. War, of course, represents an especially sensitive issue for a nation, and many claims are made regarding the "patriotic" or "unpatriotic" nature of coverage (with many news organizations desiring to not be put in the latter category). But there was more to it than just a sense of patriotism. As Kumar (2006) notes,

> In addition to cooperation between media and political elites, another element that has allowed the emergence of the current system of war propaganda arises from the structural limitations of the corporate media system. The pressure to increase profit, felt quite acutely by giant media conglomerates, has led to methods of operation that have compromised journalist ethics.
> *(pp. 51–52)*

Kumar's voice is, of course, only one in a chorus of scholars who lament corporate control and commercial interests of the press. Kellner (2004) contends that commercial media have lost their critical function that is integral to democracy and public service:

> My argument is that once the corporate media have surrendered their responsibility to serve the public and provide a forum for democratic debate...they have largely promoted the growth of corporate and state power and undermined democracy. This results in mainstream corporate press...becoming arms of conservative and corporate interests which advance state and corporate agendas.
>
> *(p. 31)*

Professional journalism, an integral part of *deep media*, also played a part in the coverage of #NODAPL, especially in the U.S., where both the *New York Times* and *Bismarck Tribune* created the kind of "he said, she said" coverage that constantly pitted Standing Rock protesters against law enforcement, the pipeline, and the state of North Dakota. As noted earlier in this book, the veneer of impartiality, which is the hallmark of professional journalism, serves to mask the corporate interests that undergird the media system. As McChesney (2004, p. 73) claims, professionalism in journalism also hides the "compromises with authority" that news outlets consistently make – ones that ultimately compromise the coverage of important issues.

The media, then, can be seen to have a "structured relationship to power," which has the

> effect of making them play a crucial but secondary role in reproducing the definitions of those who have privileged access...from this point of view, in the moment of news production, the media stand in a position of structured subordination to the primary definers.
>
> *(Hall et al., 1978, p. 59)*

These comments speak directly to the deep, institutionalized vulnerabilities in the news media industry where, in the case of the media coverage of the Standing Rock protests, the 'primary definers' would be large interests like the government of North Dakota and the corporate Energy Transfer Partners. In the case of the *Bismarck Tribune* and *Calgary Herald* coverage of #NODAPL, however, it is more complicated. While it certainly can be true that a large corporation like Energy Transfer Partners might exert an influence on coverage, the region that the newspaper represents is heavily invested in oil on numerous levels. As a result, the uncritical stance taken by some of the local newspapers on the pipeline construction is not due solely to direct influence of powerful corporations but also to the fact that local economies in "oil-patch" regions are reliant on oil revenues, with the pipeline being the infrastructure that makes economic growth in the oil industry possible. In North Dakota, there are numerous investments and livelihoods based in transporting fracked oil to markets, and it is difficult to imagine a situation in which a local paper would not feel pressure to support an industry so integral to the local and state economy.

Returning to the idea of framing, portraying the Standing Rock protesters as criminals who are a threat to social order ultimately serves economic interests, for why should criminals who threaten society deserve to be heard? Inherent in the law and order approach is "the tendency to 'criminalise' every threat to a disciplined social order, and to 'legalise'... every means of containment" (Hall et al., 1978, p. 288). Thus, the law and order framing by the papers serves to marginalize the protesters and make their motives seem suspect, which ultimately – in the case of the Standing Rock demonstrations – supports the completion and operation of the pipeline. Perhaps needless to state, when the focus is placed on law and order, environmental justice concerns are swept aside and/ or are minimized. Speaking to issues of power and mediated representations of crime, it is clear that "what hegemony ultimately secures is the long-term social conditions for the continuing reproduction of capital" (Hall et al., 1978, p. 218).

In all, seeing the law and order media coverage through the lens of deeper economic structures is not only useful but necessary to understand the system: "Although they (the courts, police, and the media) are crucial actors in the drama... they, too, are acting out of a script which they do not write" (Hall et al., 1978, p. 52). What is meant by this is that the "script," which itself is tied to the reproduction of dominant ideologies, "is the result of a set of structural imperatives, not of an open conspiracy with those in powerful positions" (p. 60). This perspective returns us to the conception of *deep media* woven into this book, for it appears that the law and order approach to coverage of the protesters has less to do with an outright conspiracy than it does to the economic structure, corporate ties, drive for profit, and professional norms of the press.

Connecting the dots: why media coverage of Standing Rock matters

Boykoff (2006) observes that "while the state is a crucial component in [the suppression of dissent], so are the mass media, especially since most people learn what they know about social movements through mass-media sources" (p. 7). After the mass protests had died down at Standing Rock, in the late spring the editorial board of the *Bismarck Tribune* published a piece about the imprisonment of Dakota Access protesters. Titled "Jail Offers Chances for All Involved," the board identified what it saw as the benefits of a new detention center, including the reduction in state by not having to transport prisoners from the old, overflowing prison:

> last year it cost Burleigh County more than $1.125 million to ship up to 50 inmates nightly to jails around the state. Morton County saw an increase in prisoners because of the Dakota Access Pipeline protests over the last few months.

In so stating, the paper made the connection between the criminality of the Standing Rock "water protectors" and the economic burden on the state. Prisons can indeed offer "chances" to those incarcerated through various programs but, despite the optimistic outlook of the newspaper, the detriments of incarceration far outweigh the opportunities, especially for people of color. As Tonry (2011) puts it, "a dose of prison can damage anyone, and usually does" (in Pickett & Ryon, 2017, p. 2). The *Bismarck Tribune* editorial, along with others published in the paper over the course of its coverage of the Standing Rock movement, suggests an alignment with the state and Big Oil rather than a critical stance towards those in power.

There is more. In his carceral research focused on framing, Gandy (2014, p. 73) claims that:

> The current orientation of mainstream journalists toward the [criminal justice system] will clearly have to be determined as part of the background research required by a strategic framing initiative. But it will not be enough to understand what the mass of journalists believe; advocates will also have to understand the nature of the enablements and constraints within which journalists actually produce the news.

Scholars like Gandy (2014), Pickett and Ryon (2017), Platt (1994), and Alexander (2010) have made compelling arguments regarding the racial disparity in incarceration in the U.S. and what it means for our society. When it comes to the racial element of certain frames Gandy (2014, p. 66) contends that "I don't believe that we have a sufficiently compelling story or narrative yet about how the massive racial, ethnic, and spatial disparities in exposure to the carceral system came to be."

Media coverage of people of color: reporting on police, courts, and crime

In their book *MediaMaking*, Grossberg et al. (2005) define "re-presentation" as an active process: "to take the original, mediate it, and 'play it back'" (p. 195). In this view, to re-present something is to change the meaning, to imbue it with ideology, because it has fundamentally changed from the original. This, then, gets at the heart of the issue surrounding media coverage of the Standing Rock protests, for it is clear that while newspapers like the *Bismarck Tribune, Calgary Herald,* and *National Post* chose to cover the issue through the lens of "lawless" protesters and the need for increased police involvement, the Standing Rock Sioux Tribe and those who supported them saw a clear and pressing environmental justice issue. Portraying those involved with the resistance to the pipeline as inherently dangerous, as a threat to social order, as a burden to the state, or as radical out-of-towners contradicts the stated concerns of those who joined the Tribe in their struggle: to protect their water source from oil contamination.

Stuart Hall's (1997) conception of the "politics of representation" complements the idea of "re-presentation" because it is clear that depictions of people of color have power. Mediated portrayals, especially to audiences who don't have much firsthand experience with members of different groups, exert an influence in terms of how the public gets to "know" people of color and how opinions are formed (Entman & Rojecki, 2000). The knowledge and opinions that audiences form from the media are essential when it comes to the public action required for substantive social change. As Boykoff (2006) puts it: "while the state is a crucial component in [the suppression of dissent], so are the mass media, especially since most people learn what they know about social movements through mass-media sources" (p. 7). Unfortunately, "the stories that we do have are unlikely to succeed in mobilizing a broad mass of the public to demand the selection of one solution, policy, or strategy over another" (Gandy, 2014, p. 66).

Ultimately, it is important to recognize that "The current law-and-order campaign is not simply another ephemeral storm of outrage that will gradually dissipate. It is part of a profound shift taking place in state power" (Platt, 1994, p. 5). Seeing people of color arrested and making their way through the court system through the lens of mainstream media is so common at this juncture that the numerous observations (Dixon & Williams, 2015[10]; Alexander, 2010; Dixon, 2008; Stabile, 2006; Rome, 2004; Hall et al., 1978) need not be repeated at length here, except to note that the mediated depictions speak to broader trends in the incarceration of people of color. Michelle Alexander (2010), in discussing her own journey to understand the racial caste system, observes that "I came to see that mass incarceration in the United States had, in fact, emerged as a stunningly comprehensive and well disguised system of racialized social control that functions in a manner strikingly similar to Jim Crow" (p. 4).

Perhaps it is possible to consider the *Economic* and *Law and Order* frames embraced by some of the mainstream news outlets covering the Indigenous-led #NODAPL movement as a part of this caste system, inextricably woven into structures of racial oppression. Pickett and Ryon (2017) likely would agree, for they find compelling evidence that "race influences the resonance of criminal injustice frames" (p. 1). As such, it is very difficult if not impossible to separate the racial identity of the "water protectors" from the focus on the need for social order evident in some of the coverage. As noted in the introduction to this book, many have perceived striking similarities when it came to the use of force in the attempts to suppress both the protests from the civil rights and Standing Rock movements. The characterization of civil rights' protest "tactics as criminal and… indicative of a breakdown of law and order" (Alexander, 2010, p. 40) can be discerned directly in the *Law and Order* framing of the #NODAPL protests by the *Bismarck Tribune* and other outlets.

Returning one final time in this chapter to the conception of *deep media* that structures this book, it is clear that there was no open, easily discernible conspiracy by the *Bismarck Tribune* to criminalize the Indigenous people who joined the resistance to the Dakota Access Pipeline. Instead, it is more useful to understand how their *Law and Order* and *Economic* framing of the protests is tethered to deeper structural flaws of the for-profit, professional media system, where the paucity of coverage of social movements is a result of fairly rigid routines and roles inherent in the media system (Boykoff, 2006). Dixon and Williams (2015) speak directly to this in their discussion of the "guard dog" theory of media reporting on people of color:

> The guard dog perspective contains aspects of both the constraints of journalism and unconscious bias perspectives. It argues that the media behave like a "sentry" for the power structure within society. This perspective argues that news stories get greater attention if they identify a phenomenon as an intruder or threat (e.g., Muslim extremists threaten national security). According to this perspective, those with the least power in the system (e.g., immigrants) receive the most bias in news coverage.
>
> *(p. 34)*

Seen from this perspective, it is useful to recognize that although journalists may have an "unconscious bias" in relation to people of color (Dixon & Williams, 2015), the most powerful factor influencing coverage of the Standing Rock Sioux Tribe lay not with individual biases of individual reporters and editors, but the rigid structures of the commercial media system that both represents and fosters institutionalized racism. It is to the Tribe that we turn in the next chapter, for it is one thing to analyze media coverage from a distance; it is quite another to learn directly from those involved about their feelings, perspectives, and knowledge.

Notes

1 The prepared, written portion of Sessions' speech was published by the U.S. Department of Justice, whose notes are available here: https://www.justice.gov/opa/speech/attorney-general-sessions-delivers-remarks-national-sheriffs-association

2 It is not often remembered that U.S. politician Reagan had starred in a 1953 Western titled *Law and Order* about a marshal who needs to leave his love to restore order in a small town.

3 Hall and colleagues, in *Policing the Crisis*, focus mostly on the adoption of law and order approaches in the U.K., but their observations are directly relevant here.

4 Flamm in particular notes that the blind pig contained, on that particular night, Vietnam war veterans who had returned home, causing even more tension with police.

5 Hall et al. (1974) provide a compelling discussion of how both the law and order approach used by Nixon and the frustrations voice by communities of color in the late 1960s and early 1970s migrated to Great Britain. Although they find that British politicians did not adopt Nixon's stance as strongly, the influence of Nixonian political stances on street crime and protests were echoed by prominent politicians. As they note, "What had really united the Conservative Party in the pre-election period was less the rhetoric of disorder, but rather a more traditionally-phrased emphasis on 'the need to stand firm,' not to give in, to restore *authority* to government" (p. 279, *emphasis in original*).

6 In particular Blumenthal points to Nixon's "war on drugs," which adopted a law and order approach to try to solve the perceived problem of marijuana use. This tactic, as Blumenthal recognizes, wasn't very popular for this particular campaign because it (perhaps unwittingly) divided families between the older, more socially conservative parents and their children who grew up in a time that was more accepting of marijuana usage.

7 Available at https://www.whitehouse.gov/presidential-actions/presidential-executive-order-task-force-crime-reduction-public-safety/

8 Inglehart and Norris (2017) note in particular the relationship between economic security and insecurity: "Growing up taking survival for granted makes people more open to new ideas and more tolerant of outgroups. Insecurity has the opposite effect, stimulating an Authoritarian Reflex in which people close ranks behind strong leaders, with strong in-group solidarity, rejection of outsiders, and rigid conformity to group norms" (p. 443). They note in particular that when the majority of economic gains go to "those at the top," economic insecurity rises, and voters put their confidence into populist authoritarian leaders.

9 Available at the Oklahoma Attorney General's archived website: https://web.archive.org/web/20170108114336/https://www.ok.gov/oag/Media/About_the_AG/

10 Although Dixon and Williams' study did address the overrepresentation of Latinos as undocumented immigrants and Muslims as terrorists, they also found that African Americans tended to be underrepresented (either as victims or perpetrators) across the board on news programs.

References

Adams, J. (1856). *The works of John Adams, second president of the United States, vol. 1.* Boston, MA: Little Brown and Company.

Alexander, M. (2010). *The new Jim Crow: Mass incarceration in the age of color blindness.* New York: The New Press.

Bernat, F. (2017). Immigration and crime. In H. Pontell (Ed.), *Oxford research encyclopedias: Criminology and criminal justice* (pp. 1–28). Oxford: Oxford University Press.

Blumenthal, S. E. (2016). Nixon's marijuana problem: Youth politics and "law and order," 1968–72. *The Sixties*, 1–28. 10.1080/17541328.2015.1115190.

Boburg, S. (2016, July 25). Donald Trump's long history of clashes with Native Americans. *Washington Post.*

Boykoff, J. (2006). *The suppression of dissent: How the state and mass media squelch USAmerican social movements.* New York: Routledge.

Brown, A., Parrish, W., & Speri, A. (2017, September 29). Police used private security aircraft for surveillance in standing rock no-fly zone. *The Intercept.* Retrieved from https://theintercept.com/2017/09/29/standing-rock-dakota-access-pipeline-dapl-no-fly-zone-drones-tigerswan/

Cohen, M. A. (2016). *American maelstrom.* Oxford: Oxford University Press.

Dixon, T. L. (2008). Crime news and racialized beliefs: Understanding the relationship between local news viewing and perceptions of African Americans and crime. *Journal of Communication, 58*(1), 106–125. 10.1111/j.1460-2466.2007.00376.x.

Dixon, T., & Williams, C. (2015). The changing misrepresentation of race and crime on network and cable news. *Journal of Communication, 65*(1), 24–39. 10.1111/jcom.12133.

Entman, R. & Rojecki, A. (2000). *The black image in the white mind: Media and race in America.* Chicago, IL: University of Chicago Press.

Flamm, M. W. (2005). *Law and order: Street crime, civil unrest, and the crisis of liberalism in the 1960s.* New York: Columbia University Press.

Friedman, D. D. (2000). *Law's order: What economics has to do with law and why it matters.* Princeton, NJ: Princeton University Press.

Gandy, O. H. (2014). Choosing the points of entry: Strategic framing and the problem of hyperincarceration. *Atlantic Journal of Communication, 22*(1), 61–80. 10.1080/15456870.2014.859977.

Goodman, A. (2006). *Static: Government liars, media cheerleaders, and the people who fight back.* New York: Hyperion.

Gottschalk, M. (2006). *The prison and the gallows.* Cambridge: Cambridge University Press.

Gottschalk, M. (2015). *Caught: The prison state and the lockdown of American politics.* Princeton, NJ: Princeton University Press.

Graham, D. (2016, July 12). The shaky basis for trump's 'law and order' campaign. *The Atlantic.* Retrieved from https://www.theatlantic.com/politics/archive/2016/07/trump-law-and-order/490940/

Gross, K. N. (2015). African American women, mass incarceration, and the politics of protection. *Journal of American History, 102*(1), 25–33.

Grossberg, L., Wartella, E., Whitney, C., & Wise, J. (2005). *Mediamaking: Mass media in a popular culture* (2nd ed.). Thousand Oaks, CA: Sage Publications.

Hall, S. (1997). *The spectacle of the "Other".* London: Sage.

Hall, S., Critcher, C., Jefferson, T., Clarke, J., & Roberts, B. (1978). *Policing the crisis: Mugging, the state, and law and order.* New York: Holmes & Meier.

Harden, C. (2017, February 27). How 3 of Donald Trump's executive orders target communities of color. *Time.* Retrieved from http://time.com/4679727/donald-trump-executive-orders-police/

Inglehart, R., & Norris, P. (2017). Trump and the populist authoritarian parties: The silent revolution in reverse. *Perspectives on Politics, 15*(2), 443. 10.1017/S1537592717000111.

Jefferson, T. (1829). In T. J. Randolph (Ed.), *Memoir, correspondence, and miscellanies, from the papers of Thomas Jefferson* (Volume III ed.). Charlottesville, VA: F. Carr, and Co.

Johnson, T. A. (1968, August 13). Negro leaders see bias in call of Nixon for "law and order". *New York Times,* 27.

Kellner, D. (2004). The media and the crisis of democracy in the age of Bush. *Communication and Critical/Cultural Studies, 1*(1), 29–58. 10.1080/1479142042000180917.

King, M. L. (2000). *Why we can't wait.* New York: Signet Classics.

Kranish, M. (2016). In M. Fisher (Ed.), *Trump revealed: An American journey of ambition, ego, money, and power.* New York: Scribner.

Kumar, D. (2006). Media, war, and propaganda: Strategies of information management during the 2003 Iraq war. *Communication and Critical/Cultural Studies, 3*(1), 48–69. 10.1080/14791420500505650.

Lee, N. (2016). Local debate viewers react to Trump's "law & order" assertion. *Philadelphia Tribune.*

Mac Donald, H. (2016). *The war on cops: How the new attack on law and order makes everyone less safe.* San Francisco, CA: Encounter Books.

McChesney, R. W. (2004). *The problem of the media: U.S. communication politics in the twenty-first century.* New York: Monthly Review Press.

Minot, G., & Sanger, G. (Eds.). (1863). *Statutes at large and treaties of the United States of America, from December 3, 1855 to March 3, 1859.* Boston, MA: Little, Brown and Company.

Nelson, L. (2016). Trump: "I am the law and order candidate." *Politico.* Retrieved from https://www.politico.com/story/2016/07/trump-law-order-candidate-225372

Nicholson, B. (2017, May 30). Dakota access pipeline, law officers had close relationship. *AP Worldstream*. Retrieved from https://search.proquest.com/docview/1903683443

Nunberg, G. (2016, Jul 28). Is Trump's call for "law and order" a coded racial message? *National Public Radio*. Retrieved from https://search.proquest.com/docview/1807 409764

O'Donnell, L. (2017). *Playing with fire: The 1968 election and the transformation of American politics*. New York: Penguin Press.

Panetta, A. (2016). Trump's big speech will borrow Nixon's law-and-order theme in echo of '68. *The Chronicle Herald*. Retrieved from http://thechronicleherald.ca/ world/1381197-trumps-big-speech-will-borrow-nixons-law-and-order-theme-in-echo-of-68

Pickett, J., & Ryon, S. (2017). Race, criminal injustice frames, and the legitimation of carceral inequality as a social problem. *Du Bois Review: Social Science Research on Race*, *14*, 1–26.

Platt, A. (1994). The politics of law and order. *Social Justice*, *21*, 3–13.

Rome, D. (2004). *Black demons: The media's depiction of the African American male criminal stereotype*. Westport, CT: Praeger.

Sacco, V. F. (2005). *When crime waves*. Thousand Oaks, CA: SAGE Publications.

Stabile, C. A. (2006). *White victims, black villains: Gender, race and crime news in US culture*. London: Routledge.

Steffens, L. (1928). How I made a crime wave. *The Bookman; a Review of Books and Life (1895–1933)*, *68*(4), 416.

Tonry, M. (2011). *Punishing race: A continuing American dilemma*. New York: Oxford University Press.

Vasquez, M. (2018, February 12). Sessions invokes 'Anglo American' heritage of sheriff's office. CNN. Retrieved from https://www.cnn.com/2018/02/12/politics/jeff-sessions-anglo-american-law-enforcement/index.html

7

INDIGENOUS PERSPECTIVES ON THE DAKOTA ACCESS PIPELINE, POLITICS, AND THE MEDIA

The Standing Rock Sioux Tribe and journalists speak

Of all the chapters in this book, this one was the one I was simultaneously the most eager to write and the one I dreaded writing the most. I was looking forward to describing the thoughts and perceptions of the Standing Rock Sioux Tribe in their own words, but I feared it for the simple reason that I didn't want to get the story wrong. If their long and varied history and the events in 2016 and 2017 at Standing Rock did not already make it clear, their story is one worth telling and they are a people worth knowing. In this chapter I have attempted to accurately represent their history and also their words, emotions, and perspectives on many issues, including the media and the Dakota Access Pipeline. While I do not speak for the Tribe, I have carefully attempted to reproduce their words and thoughts here while providing background and context.

In late May I traveled from Seattle, Washington to Bismarck, North Dakota to conduct a series of long-awaited, in-depth interviews with members of the Standing Rock Sioux Tribe as well as experienced journalist, editor, and scholar Mark Trahant, whose work I have followed for decades. The purpose of the trip was clear: I knew that this book, exploring how the news media covered the events at Standing Rock in 2016 and 2017, would not be complete without their perspective. It is one thing to study, from a distance, how the media cover an issue; it is quite another thing to understand firsthand how the people who lived (and continue to live) through the issue think and feel about it.

On the plane I sat next to a young white woman, a mother of three who grew up in North Dakota but now lived in Boulder County. As she cradled her baby in her lap, she shared with me that, as in North Dakota, fracking had come to Colorado. While fracking was happening in the county next door, she and her neighbors in Boulder County were preparing to fight it coming to them: "Why would we want our water taken and the ground polluted with chemicals?" Having said this, she saw the arrival of fracking in her county as nearly inevitable due to the immense power of oil companies. "They have a formula," she explained. "If the people or the city don't want it, they tie them up in court battles until there's no more money left. Once the city is bankrupt, then they come in and fracking happens anyway."

The purpose of including her story here is not to reframe the story of the Standing Rock resistance to the Dakota Access Pipeline through the lens of middle-class white concerns. That is, the story of the Indigenous-led #NODAPL movement does not take on greater importance because predominantly white communities are increasingly impacted by fossil fuel development. Instead, I recount her story for one reason: it directly ties into a key point made by members of the Tribe about the power of Big Oil and its impact on the people and environment everywhere. As such, this chapter represents a return to the assertion made in the first pages of this book: namely, that while the events at Standing Rock in 2016–2017 represent a standalone issue worthy of deep exploration, it also is true that the study of those events reveals much about deeply entrenched power structures, about tensions over fossil fuels at a time of increased concerns about climate change, and about the need to recognize other ways of thinking about environmental issues and human rights.

Qualitative inquiry: the critical interpretive framework of interviews

As noted in the introduction to this book, the qualitative research that comprises the foundation of this chapter (as with the book as a whole) is inductive, applied, deeply critical, and engaged – four elements that are inextricably and intentionally intertwined here. At this juncture in the research it is worth describing why I chose interviews over other qualitative methods. As Lindlof and Taylor (2019, p. 219) note, we live in an "interview society," where the format functions as a way to constantly establish and formulate public identity. There is a certain comfort and familiarity with the idea of interviews because of their near ubiquity in contemporary society. But interviews are about more than a simple question-and-answer session and thus require a thoughtful approach, for there are many issues at stake when it comes to conducting interviews with a sense of care for those studied. So, why interview the Standing Rock Sioux Tribe? One commonly held reason for doing interviews is the idea that they generate "factual information about the world" or about the event or people being studied (Lindlof & Taylor, 2019, p. 222). Instead of focusing on the rather positivist language of "fact gathering," however, it might be more accurate to note that the goal of interviews is to reveal what people know, what they have experienced, and what their perspectives are (Lindlof & Taylor, 2019), which means this research is less focused on the facts of what happened and more focused on revealing perspective and context. From this view, in-depth interviews permit researchers to elicit 'deep' answers to their questions about "polarizing, sensitive, confidential, or highly personal topics" (Guest et al., 2013. p. 117).

Along these lines, the interviews conducted with the Tribe closely align with the *interpretivist* approach, which recognizes the reductionist pitfalls[1] inherent in the scientific approach "that often misses the point of qualitative research. Instead, this approach…is more interested in interpreting deeper meaning in discourses that is represented in a collection of personal narratives or observed behaviors and activities" (Guest et al., 2013, p. 5). In a very real sense, *interpretivism* gives space for elicitation of the deep "webs of significance" and contextualization that Geertz (1973, p. 311) identified as a goal of engaging in *thick description*.[2] Thus, the interviews as described here place focus on "revealing multiple realities as opposed to searching for one objective reality" (Guest et al., 2013, p. 6). Although Geertz (1973) writes from the specific perspective of an anthropologist, what he describes can be easily broadened to the qualitative researcher in general: namely, "one who 'inscribes' social discourse, *he writes it down*. In so doing, he turns it from a passing event, which exists only it its moment of occurrence, into an account, which exists in its inscriptions and can be reconsulted" (p. 317). In other words, *thick description* is not simply a dry chronicling of an event, worldview, or experience but involves context so as not to "divorce it from what happens" (p. 317).

Lindlof and Taylor (2019, pp. 11–13) identify a few key characteristics of the interpretivist approach, which I have condensed and summarized here.[3] First, it is inherently *inductive*, where theoretical assumptions are built "from the ground up" through research, and not the reverse. It also is *iterative*, with constant revisitations and reformulations as the research progresses. Related to this is that interpretivism often is 'expansionistic,' helping us to understand broader phenomena within and outside of the avenues of a specific area of research. Importantly, this perspective interrogates the role of the researcher in the research, and seeks "to achieve *deep, empathetic understanding* of the human actions, motives, and feelings." Towards this end, researchers should adopt an *emic* perspective by preserving "the subjective experience of social actors by depicting their communication as something that is '*distinctively meaningful for them*'" (p. 12, emphasis in original).

The adherence to many principles in the interpretivist approach means that, for these interviews and this chapter in particular, this research recognizes the need to dismantle "academic hegemony" through the recognition of six principles rooted in Indigenous perspectives as identified by Susan Miller and James Riding (2011, p. 3) in *Native Historians Write Back: Decolonizing American Indian History*:

- First, Indian sovereignty derives from inherent powers that predate the U.S. Constitution.
- Second, the lands and resources in what now constitutes the United States passed from Indian to non-Indian hands through serial acts of duplicity, violence, deceit, and coercion.
- Third, European claims to lands belonging to others by virtue of discovery are rooted in racially based assumptions and articulated in a language that characterizes Indians as inferior savages who lack fundamental rights accorded to "civilized" peoples.

- Fourth, the invaders used this language of racism to rationalize their aggression against unoffending Indians.
- Fifth, those nineteenth-century discourses of colonialism are entrenched in contemporary academic and legal thought.
- Sixth, colonialism must be seen for what it is: a crime against humanity (p. 3).

This research takes these principles to be true, and employs them to ground the work in a deeply critical perspective that recognizes the need to deconstruct existing power structures, including those within academia, as a way to encourage action and engagement in the world.

Working with the Standing Rock Sioux Tribe as a sovereign nation

Qualitative researchers Guest et al. (2013, p. 1) contend that "there is no right or wrong way of conducting a qualitative research project." Although I take their central point that there are many different methodological approaches to gather information in qualitative inquiry,[4] when it comes to working with groups that have been historically marginalized, there is quite a lot that can be done "wrong" in terms of harm (intended or unintended) to the group in question. That is, while many different methods may be appropriate for the qualitative researcher to gain desired information, the *way* in which these methods are implemented can have consequences for the Indigenous partners in a research project. As Lindlof and Taylor (2019) have observed, "if interviews are learning situations, the interviewer should be a willing, self-effacing student. If interviews are cross-cultural encounters, the interviewer should be a fluent speaker of local languages and a sensitive traveler across cultural borders" (p. 220).

In this research I use the word "partners" or "collaborators" to refer to the Tribe for the central reason that they are not my "subjects" or "participants" in this study. Instead, I see that they have chosen to engage in a partnership with me, where they have trusted me to accurately reflect their perspectives, experiences, and stories, and where we generate knowledge together. As such, this effort falls into the tradition of what Greenwood and Levin (2005) term *cogenerative inquiry*, a key component of action research that "aims to solve pertinent problems in a given context through democratic inquiry in which professional researchers collaborate with local stakeholders to seek and enact solutions *to problems of major importance to the stakeholders*" (p. 54, emphasis added). In this vein, then, my work with the Tribe is indeed "based on bringing the diverse bases of their knowledge and their distinctive social locations to bear on a problem collaboratively" (p. 54) with a clear eye for improvement of the media's representation of them. As such, these interviews are centered less on a "researcher–subject" dichotomy than a partnership where I have attempted to accurately represent their viewpoint, perspectives, and stories.

As a researcher, I wanted to avoid the "helicopter" research discussed by many scholars, where researchers (literally or figuratively) drop into an area of study, do their research on a particular group, and then leave without further contact. When it comes to Indigenous groups like various First Nations tribes in Canada, Bharadwaj (2014) notes that "the traditional 'helicopter' approach to research applied in communities has led to disenchantment on the part of First Nations people and has impeded their willingness to participate in research" (p. 15). While researchers thus would want to avoid this for reasons based on self-interest (you don't want to alienate various groups because that precludes finishing your work), this narrow focus on this consequence precludes understanding the full impact. Bharadwaj continues:

> In this arguably failed approach, researchers develop projects without community input, collect data without the full knowledge and consent of participants, seldom share findings, and almost never create mechanisms to continue successful research projects or programs that would greatly inform public health and environmental policy development.
>
> *(p. 15)*

For engaged and action-oriented scholarship, being able to employ the results of one's research to further social justice aims is the clear goal, for the work can and should have consequences in the real world.

Towards this end, I have sent the Standing Rock Sioux Tribe all of my research for this book. As part of a continuing dialogue with them (one I hope to continue long after the book has ended) I have offered them the chance to correct any errors I have made or alter any of the ways in which they are represented in this book. As a researcher, being flexible, open, and willing to learn has become a key component of my methodology. And I *have* learned: in the interviews I came to understand much more about the history of the Tribe than can be gained from just reading scholarly works. I also learned how they view key moments in their history leading up to the present that are so central to understanding their current perspectives on the Dakota Access Pipeline. As I did the interviews, it became clear to me that, in some instances, I didn't have sufficiently deep knowledge to even formulate the right questions for them. It was not due to my lack of reading, but perhaps that's the point: some things you don't learn about from books alone. It was only due to their patience with me that I learned as much as I did throughout the process and I am grateful that they chose to collaborate with me.

Interviews: the people and methods

As noted earlier, I interviewed several members of the Standing Rock Sioux Tribe as well as the journalist and scholar Mark Trahant. I chose the Tribe due to their prominence in the #NODAPL movement: they began and ended as the centerpoint of this book and the reason I chose this research project. I asked to interview Mark Trahant because his lengthy experience as a journalist and editor writing about all types of issues, including those related to Indigeneity, meant that he could contextualize what happened at Standing Rock in 2016 and 2017 through a broad lens relating to history, politics, and the news media industry.

Interview style and format

The type of question-and-answer style I employed to interview both journalist Mark Trahant and the Standing Rock Sioux Tribe was semi-structured, which Rowley et al. (2012) define as a series of open-ended questions where "the interviewer has a series of questions or topics to be covered, but the interviewee has a great deal of leeway in terms of how to reply" (p. 95). One of the benefits of semi-structured interviews is that they typically "incorporate both open-ended and more theoretically driven questions, eliciting data grounded in the experience of the participant" (Galleta, 2013, p. 45) and they usually reveal "richer and more insightful data" (Rowley et al., 2012, p. 95) than other types of interviews.

Thus, while I entered the interviews with a set list of prepared questions (see Appendices A and B), the questions were designed to elicit open-ended answers and there was much flexibility in terms of where the interviews would go from there. Most times my interviewees' responses would prompt follow-up questions from me or requests for clarification. Interviews with the Tribe went on much longer than I had anticipated, and I believe that is due to the amount of information in their answers and also their graceful patience with my many questions.

Once the Tribe agreed to partner with me in the study, I respected their choice as to which members would be interviewed. While I made gentle suggestions (I wanted the chance to interview both men and women and I wanted to speak to some tribal members who were relatively active or involved in the events of 2016 and 2017), I let them make the choice as to who exactly I would meet. In the end, I deeply appreciated their choices, which included current Standing Rock Tribal Chairman Mike Faith, Councilmen Charles Walker and Courtney Yellow Fat, Councilman Brandon Mauai, and Phyllis Young.

A note in is order about the format of the interviews, as typically they are one-on-one. With Mark Trahant this was the case, but with the Tribe it was different. I had anticipated interviewing each tribal member separately; when I arrived, however, it was a more group-oriented, communal, and fluid style. I initially sat with Chairman Faith, Councilman Walker, and Councilman Mauai, with each answering my questions in turn and when they felt they had something they wanted to say. Some members would then leave, and others would arrive, including Councilmen Walker and Yellow Fat. So, while I had anticipated interviewing three tribal members, when all was said and done I had interviewed five. Only Phyllis Young and I sat down for a truly one-on-one interview. Not only did this fluidity of interaction not concern me, but the more "focus group" style of interview seemed more natural and relaxed. Focus group-style interviews exist as both a way to obtain information from research partners and to gauge interactions between them, and I include my observations on those in the pages that follow.

Background of the Standing Rock Reservation and Standing Rock Sioux Tribe

The history of the *Oceti Sakowin*, also known as the *Seven Council Fires* (or, more colloquially, the *Great Sioux Nation*), is lengthy. For reasons related to space and the accompanying desire to not recreate the good work already published on this history, some discussion is provided in this section of the Standing Rock Sioux Tribe and the Standing Rock Reservation – understanding that the histories of both are intimately intertwined. In addition, significantly more history and background on the Tribe is woven into the interviews chronicled below. For those readers interested in more lengthy histories and background, I recommend Michael Lawson's *Dammed Indians Revisited* (2009), Vine Deloria's *Custer Died for Your Sins* (1988), Richard Erdoes' *Native Americans: the Sioux* (1982), and Guy Gibbon's *The Sioux: the Dakota and Lakota Nations* (2003).

Gibbon (2003) describes the Sioux as "a loose alliance of tribes in the northern plains and prairies of North America. They speak Siouan, a linguistic family that at contact was among the most commonly spoken language stocks north of Mexico" (p. 2). Providing a brief history of the reservation itself, Sprague and Sprague (2015, p. 1) note that

> The Standing Rock Reservation comprises 2.8 million acres and occupies territory in two states, North Dakota and South Dakota.... On April 29, 1868, the boundaries of the reservation were described under the Treaty of Fort Laramie.... The Executive Order of March 16, 1875, extended the reservation's northern border to the Cannonball River. However, in 1889, Congress reduced the Great Sioux Reservation, dividing it into six separate reservations, including the Standing Rock Sioux Reservation.

The Standing Rock Reservation rests along the western bank of the Missouri River within the larger Missouri Basin. Its members include Pabaksa and Ihunktuwona bands of the Dakota "Oyate" ("People"), and Hunkpapa and Sihasapa bands of the Lakota Oyate (Eagle, 2016). The Tribe is a member tribe of the *Oceti Sakowin*. Gibbon (2003, p. 2) notes that the "Sioux alliance of tribes has three main divisions, the Dakota to the east, the Yankton-Yanktonai in the middle, and the Lakota to the west, with the latter now more numerous than the others combined." In addition, the seven main subdivisions, including the "Mdewakanton, Wahpekute, Sisseton, Wahpeton, Yankton, Yanktonai, and Lakota – are recognized by the Sioux as ancestral political units, the *Seven Council Fires* (*Oceti Sakowin*), whose origins extend back to their homeland in the present state of Minnesota" (Gibbon, 2003, pp. 2–3).

The Standing Rock reservation was created as the direct result of the U.S. policy that concentrated the Sioux into ever-smaller "reserved areas." This meant that "by 1889 the separate tribes of the Sioux Nation had found themselves confined within the shrunken boundaries of their present reservation" (Lawson, 2009, p. 28). Lawson (2009) notes that "the establishment of separate reservations blurred longstanding tribal distinctions [and] tribal members soon began to identify themselves according to their reservations. Hunkpapa descendants, for example, began referring to themselves as the Standing Rock Sioux" (p. 29).

Interviews with the Standing Rock Sioux Tribe

On April 27, 2018, I drove from Fargo, North Dakota to the Standing Rock Reservation, passing through the town of Bismarck. The drive was relatively lengthy – about four hours – and I got to see much of the flat, agricultural landscape comprising the state from the vantage point of eastbound Highway 94. Early in the drive I had turned on the radio and heard, almost immediately, a song that was new and surprising to me. The song was about a wife who was tired of her husband cheating on her, and included such lyrics as:

> Well you're leaving me at home to keep the tee pee clean/ Six papooses to break and wean/ Well, your Squaw is on the warpath tonight./ Well I found out, a-big brave chief/ The game you were hunting for ain't beef/ This war dance I'm doing means I'm fighting mad/ Well that fire water that a you've been drinking/ Makes you feel bigger but chief you're shrinking/ Now don't hand me that old peace pipe/ There ain't no pipe can settle this fight/ Your Squaw is on the warpath tonight.

The song was by country singer Loretta Lynn[5] and employed so many stereotypes about Native Americans that I almost drove off the road listening to it. While listening to the echoes of the song in my head I noted that every state highway sign was marked with a stylized profile of an "Indian Chief" with a long feathered headdress. Even the name of the state – "North Dakota" – was a reference to the Sioux.

I discuss these instances of what I saw not to engage in debate about whether or not Lynn's song or the road signs, with all their facile stereotypes of Indigeneity, are offensive, but because they seemed to fit perfectly with the overall atmosphere I felt being in the state. Ella Shohat and Robert Stam (2014, p. 221) refer to this as *inferential ethnic presence*, where "racial undertones and overtones…'haunt' everyday social life" as they represent "the repressed stories, the sublimated agonies, and the buried labor of people of color" (p. 221).

My feeling regarding the permeation of the Sioux into the everyday lives of people living in North Dakota as a kind of unending, subtle background referent was only strengthened in my brief visit to Bismarck. When I stopped by the Bismarck-Mandan Convention and Visitors Bureau, the staff person was happy to help me find restaurants and coffee shops, pulling out maps to show me and providing directions. When I asked if she could tell me more about the Dakota Access Pipeline, including where it crossed the water and where the protests had occurred, her smile dropped somewhat and she waved (perhaps unintentionally) in the opposite direction of where I knew the pipeline to be. She changed the subject quickly and told me about other sightseeing I could do. In general, people in Bismarck and the nearby Mandan didn't seem to want to talk about what had happened at Standing Rock. This, perhaps, is understandable, as it was a contentious time that had seemed to paint the towns (and state) in an unfavorable light, as the *Bismarck Tribune* editorial board had made clear in several stories. But the ubiquitous references to the Sioux throughout the state seemed to make the lack of overt discussion about the Standing Rock protests stand out even more in sharp relief. Leaving the towns, I traveled south towards Fort Yates and to the reservation of Standing Rock.

Members of the Standing Rock Sioux Tribe who chose to be interviewed

Arriving to the government building North Standing Rock Avenue on the banks of the Missouri River in Fort Yates, North Dakota, I met with Danielle Finn, External Relations Officer for the Tribe, who took me into Chairman Mike Faith's office. There I met Chairman Faith and several other members of the Tribe, including Councilman at Large Charles Walker, Long Soldier District Councilman Brandon Mauai and (later in the interview), Councilman at Large Courtney Yellow Fat. After the group interview ended, I met and interviewed Phyllis Young.

Chairman Mike Faith was elected in September 2017. His election campaign focused on the increased need to help support Tribal members by improving services, including housing, elder care, education, and health care in Standing Rock. As he stated to the *Associated Press* (Nicholson, 2017), "We kind of neglected our own.... We did what we had to do, but we didn't realize we were going to hurt our economy so much." Chairman Faith has made inter-government negotiations and discussions a priority for his administration: in late May 2018 the Chairman met with U.S. Secretary of the Interior Ryan Zinke at United Tribes Technical College. The bio provided by Standing Rock reads (in part) as follows:

> Chairman Faith is from the Long Soldier District and has represented the Standing Rock Sioux Tribe as an elected councilman on Tribal Council for seventeen years prior to being elected Chairman. Chairman Faith's Lakota/Dakota name is Tatanka Akicita, which means Buffalo Soldier. Chairman Faith lives up to his name through his long connection and dedication to the Lakota/Dakota People's most sacred animal, the buffalo. Currently, Chairman Faith is the Vice President of the Intertribal Buffalo Council and the Associate Director of the North Dakota Buffalo Association. Chairman Faith sits on many other boards, but one in

particular he holds dear to his heart is the Standing Rock Community School Board, of which he has served on for over twenty years.

Tribal Councilman at Large Charles Walker has taken an active role in the administration since his election in late 2015. In late May 2018 he, along with Chairman Faith, met with U.S. Secretary of the Interior Ryan Zinke at United Tribes Technical College. Councilman Walker also serves on the Judicial Committee.

Tribal Councilman at Large Courtney Yellow Fat teaches Culture Education at Standing Rock Community School and also was the Department Chairman of Cultural Education there. He owns Buffalo King Music, is a screenwriter, and was a team leader at Lakota Thunder Drum Group. Councilman Yellow Fat was elected to the Standing Rock Council in September 2017. His brother, Councilman Dana Yellow Fat, serves with him on the Tribal Council.

Long Soldier District Councilman Brandon Mauai also was elected in September of 2017. This is information from his bio:

> Councilman Brandon Mauai was elected to the Standing Rock Sioux Tribal Council in 2017. He is a lifelong resident of the Standing Rock Sioux Reservation. Prior to his service on the Standing Rock Sioux Council, Councilman Mauai was a youth minister and Deacon with the Episcopal Church on the Standing Rock Sioux Reservation.[6]

Standing Rock Sioux Tribe member, Hunkpapa Lakota, and activist Phyllis Young was a founding member of the Women of All Red Nations (WARN) and a member for many decades of the American Indian Movement (AIM). She was active in the resistance again the Dakota Access Pipeline, forming one of the main camps. She served a four-year term on the Tribal Council at Standing Rock and is known as a "revered elder" in the Tribe. She is very active politically and has traveled extensively, speaking at conferences and participating in marches. In the capacity of Team Leader, she was awarded a grant – the "Oceti Sakowin SOLVE Fellowship" – from the Massachusetts Institute of Technology to create a tribal utility company based on solar energy. Phyllis also is a treaty writer, a fact that became particularly relevant during our interview.

The importance of context and history

During my interviews with members of the Tribe about the Dakota Access Pipeline, they referred to several events to which they ascribe a great deal of significance. Because these historical events are so central to tribe members' perspectives, during the description of the interviews I provide background on the events in the context of their statements. I mention these here not as a way to provide a definitive history of these events and developments (as they have been researched and described in much greater depth than I am able to do here) but instead to provide background to some of the Tribe's observations.

In particular, you will read that some tribal members referred to a particular "papal bull" known as the *Inter Caetera* (or, more colloquially, the *Doctrine of Discovery*), the U.S. government's Pick-Sloan Program that resulted in the Oahe dam, the hanging of 38 Dakota people after the *Great Sioux Uprising*, the theft of their sacred Black Hills after gold was discovered, and the *Indian Offenses Act*. As such, a note is in order about these interviews: the tribal members to whom I spoke patiently took me to school on many issues relating to Indigeneity, sovereignty, and history. Everything I thought I knew about Native American land rights and history paled in comparison to the deep knowledge and philosophical perspective the Tribe demonstrated in these interviews. To come up to speed as quickly as I could I read the court cases and the original *papal bull*, I watched the documentaries, including *The Doctrine of Discovery: Unmasking the Domination Code* and *Dakota 38*, and I delved into scholarly reading. Because of this, I can start to scratch the surface of the depth and complexity of the Tribe's perspective on a modern-day privation like the Dakota Access Pipeline a bit more clearly: namely, it cannot be untethered from historical and cultural context.

Group interview

As noted above, I sat down with several members of the Tribe at the large table in the Chairman's office at the Standing Rock government office. Because I wanted to show my appreciation for their time, and because they could not, as elected members, accept any of the grant money I had received to pay my research partners, I brought them gifts made by Coastal Salish artists from the Northwest. In fear of running out, I had brought perhaps too many items, and the Chairman began to gently joke that every time I opened my small carry-on suitcase that something new would come out! I admitted that it looked a bit comical but that I wanted to show my appreciation for the Tribe's time and consideration. We settled in and began the interview. The questions I had prepared for the Tribe centered mostly on their perceptions of the Dakota Access Pipeline itself as well as the media coverage of the protests that followed. Of note was that many of the tribal members present referred to the pipeline protests as the "gathering," for reasons that become clear during this interview.

The Dakota Access Pipeline, American identity, and the need to recognize sovereign rights

When I asked about their perceptions of the Dakota Access Pipeline itself, tribal members spoke about different aspects of the pipeline, yet they all came around to a central point about nationhood and sovereignty. Chairman Faith, the first one to speak on this, noted his chagrin about the way in which the pipeline was proposed:

> I'll tell you what I've been telling people: yes, we are citizens of the United States, we're citizens of North, South Dakota, but we're unique by treaty.

That's where the government-to-government consultation comes in, and a lot of that was bypassed. So all of a sudden you got a resistance. Well they didn't consult with us. They moved the pipeline from north of Bismarck down here, so things like that start building up.

Here, Chairman Faith observed that "we just want to be treated as equals" and highlighted the sovereign nature of the Standing Rock Sioux Tribe. He specifically desired government-to-government consultation "because our cultural resources were at risk [and] our environment, our clean water." Councilman Mauai agreed, noting that government-to-government negotiation would have been an important deciding factor in terms of how the Tribe would have responded to the pipeline. Similarly supportive of this idea, Councilman Yellow Fat offered the following hypothetical scenario as a way to understand what happened between the Tribe and the state of North Dakota:

> One thing about the consultation of nation-to-nation: I came up with an analogy. What would happen if you put a pipeline underneath the Red River on the east side of the state. Now that flows north, and it flows into Canada. You're not going to tell Canada, "oh, you didn't come to our meeting at the local library basement." They're going to laugh you off, you know?

Yellow Fat specifically wanted it to be known that the Tribe "needs a place at the table… to negotiate water treaties with the United States government." He also took issue with the placement of the pipeline, which moved downstream from the town of Bismarck so as not to endanger their water supply but moved upstream from the Tribe's water source:

> Originally it was north of Bismarck. So it was too close to there. If there was a break it would affect Bismarck, Mandan and their water supply. So it makes the people here and myself think, "oh so my life and my children and grandchildren's future is expendable?" Is that what you're saying about thousands of people that'll be affected south of here?

To this statement, Chairman Faith nodded and agreed: "Expendable, yes."

The deep significance of the tribal "gathering" around the pipeline

What came next regarding the Tribe's perception of the pipeline was unexpected to me. Councilman Walker shared that the significance of the "gathering" of pipeline resistance was the renewed solidarity between the Standing Rock Sioux Tribe and other bands within the Great Sioux Nation. That, he said, was the "biggest event" and the one that was not reported on by the mainstream media:

> It was fascinating because it was about 150 years since the last time that had happened. Our people had distinct bands had come together, so

it was really historical and I think that one thing that you won't hear [from] the *New York Times*, *Bismarck Tribune*, and all these different media outlets.

Councilman Walker continued by noting that, in addition to renewed solidarity within *Oceti Sakowin*, the one event that generations of Standing Rock Sioux Tribe would remember was the historic reconnection with the Crow Nation:

> The biggest event to not only Standing Rock but our Oceti Sakowin relatives wasn't the gathering itself, it was that the Crow came into our territory to ask us if they [could] provide help for us. The last time that had happened was 246 years. So that was the largest event. So if there's a Winter Count[7] somewhere that someone's keeping track of our people, it wouldn't be about a pipeline, it would be about the Crow coming here.

When I noted that I had not read about that in the media, he answered "That's because we're oral historians. That tells you that our viewpoint of the world is connected to the land and our oral history will tell a story you cannot hear from anyone on the outside looking in."

As a final note on this, Councilman Walker noted the new connections made with hundreds of other Indigenous nations, observing that, if not for the Dakota Access Pipeline controversy, "I wouldn't have met some of the people in the different coastal Salish communities or the Plateau Indians and Yakama, or Warm Springs or the Nimi'ipuu Nez Perce, or [those Tribes] down in California or even Arizona." The significance, for him, could not be overestimated, namely that

> all these nations that came, you know, they changed the world view and I feel like it's a preparation for something in the future, because we're going to have to stand together and we see it evident today. With the Trump administration…there's going to be a challenge to who we are today, even as they want to consider us a race rather than a distinct government.

According to reports that began to surface a few days prior to my interview with the Tribe,

> the Trump administration made the remarkable step of asserting that tribal citizens should be required to have a job before receiving tribal health care assistance. Several states are seeking to force the requirement on tribal health care systems that have always operated within their sovereign nations. The Trump administration contends that tribal members should be considered a race, not a political class, as courts have always viewed them, and should not be exempted from state regulations.
>
> *(Brewer, 2018)*

Thus, Councilman Walker's comments speak to the Tribe's concerns about the loss of sovereign status, which would significantly alter the Tribe's way of life, including the ability to use casinos as a modest source of much-needed revenue as well as having exemption from the Medicaid work requirement so that the Tribe can have access to health care.

Politics, presidents, and the pipeline: the need for context

Given that Councilman Walker brought up politics, and specifically Trump, I asked for his perception of Trump and the new administration as well as politics in general. His answer revealed much about the Tribe's perception of their history in relation to the United States as well as how they conceive of their future:

> It's no accident that the conditions on every reservation are the way they are. That's just the way it's set up. The Republicans, they want to cut us off and say "if you're sovereign, well, do it yourself," and the Democrats want to say, "here's a little bit of money, but you have to remain in poverty in order to utilize it." So it's not a matter of party lines: it's that we've been in survival mode since the arrival of the first settlers. We've always adapted. *We lived through Andrew Jackson, we lived through Lincoln, we'll live through Trump. So, you know, we'll always be here.*

While I understood the reference to "living through" former President Andrew Jackson (who signed the Indian Removal Act of 1830 designed to forcibly relocate American Indian tribes living in the southeast from their land to give to white settlers, among other actions), I did not initially comprehend the Tribe's negative feelings about Lincoln. But as Chairman Faith, Councilman Walker, and Councilman Mauai explained, President Lincoln chose to hang 38 Dakota men who had participated in an uprising. As Erdoes (1982, p. 11) tells it, "Thirty-eight Santee Sioux were hanged for having taken part in the Great Sioux Uprising of 1862. The uprising was caused by the government's failure to deliver to the starving Indians much-needed food supplies." Lincoln was given the option to execute all the Dakota who had taken part in the uprising (almost 300), but he chose to hang 38 which, as the Tribe told me, was "historically the largest mass execution on U.S. soil" to this day. Danielle Finn had just come in and joined the conversation, noting that "It happened the day after Christmas, all the hanging. And then [Lincoln] signed one week later the Emancipation Proclamation." Here she identifies her thoughts about a significant inconsistency: Lincoln sentencing Dakota people to death the day after a Christian holiday and a few days prior to freeing African slaves.

In a fluid manner, Chairman Faith then made a connection between the mass hanging of Dakota Sioux and the "Massacre of Wounded Knee." During this event in 1890, U.S. soldiers (including some of Custer's original 7th Calvary), killed hundreds of Sioux Chief Big Foot's people near Wounded Knee Creek. The reason for the massacre was U.S. fears about the growing popularity of the peaceful and ceremonial

"Ghost Dance" – fears that ultimately led to the "murder" (the Chairman's words) of the elder prophet, Chief, and former warrior Sitting Bull outside of his home.[8]

What I understood from this discussion, in which the Tribe moved seamlessly from describing their views of the Dakota Access Pipeline to modern politics and then to their historic treatment by the United States government and past presidents, is three-fold. First, for them the Dakota Access Pipeline simply represents the latest injustice in a very long, historic line of controversial actions taken by the U.S. government that threatens the Tribe's health and way of life. Second, tribal solidarity means a great deal not only in a political sense (where connection to Tribes increases political sway and power) but also on a deep, cultural level. As a result, they see the unwanted pipeline as paradoxically bringing a benefit they cherish: renewed connections to other Tribes from within the *Oceti Sakowin* but also around the U.S. and the world. In addition, they noted the importance they ascribed to having a connection with Black Lives Matter and the Nation of Islam. For them, the connections made at the "gathering" were valued greatly. Finally, this thread of discussion starting with the pipeline demonstrated to me their unshakable belief in their enduring resilience and their ability to survive just about anything – because they already have done so innumerable times.

The original intent of the camps in response to Dakota Access Pipeline

Early in our interview, the Tribe wanted to talk about the original intent of the camps – including *Sacred Stone* and *Oceti Sakowin* – that formed from the Tribe's concerns about the pipeline. Councilman Walker spoke of the origins of the resistance, a 500-plus-mile run that started with youth who wanted to draw attention to the issue of pipelines and oil.[9] He then noted that a meeting with about 40 Standing Rock community members was held in Long Soldier to decide how to approach any resistance to the pipeline:

> What did we do? We gathered and the first thing was prayer. [Then]... there was a ceremony... and those who went to the ceremony were asking for direction about what to do. There were specific instructions that were received from that ceremony and... the first one was to pray with our chanupa, our pipe. Next we had to unite all of our brothers and sister Tribes. Three, no violence; and four, no one was to go to jail. If we follow all these protocol, we will win. We will beat this. But we must follow this.

In this, Councilman Walker was clear that there was a good protocol established for the "gathering" and resistance to the pipeline that was to come. He then noted that when non-tribal protesters arrive to the campus later, that "they had no idea of the spiritual protocol which was in place for our people here." Rather seamlessly, he referenced Sitting Bull,[10] noting that he had a "vision of victory" for the Battle of the Greasy Grass[11] in 1876, "where his instructions were to the people to not take the spoils of war."

Chairman Faith agreed, noting that

> I keep saying that we lost control, because…the intent was trying to get the
> seven Sioux tribes back together in a good way: there was a lot of prayer,
> there was a lot of ceremonies, and when that went away, there was a lot of
> things that were coming out of [the camps] with negative impact. Drinking.
> Rapes.

He also added that while he agreed in principle with resisting the placement of the
pipeline and endangering the Tribe's water, "you know the fight's the fight, but
do it in federal court. Do it the right way without getting somebody hurt. I guess
I would kind of look at it that way." Adding to this later, Councilman Yellow Fat
reiterated that "it was supposed to be spiritually focused," adding that when non-
tribal protesters understood their ceremonies and instructions, especially those
saying not to take up arms, then they would be following the Tribe's protocol:
"But like the Chairman said, some people came from around the world, and some
of them came with their own agendas." Some of those people – but not all – "saw
it as a way to stand up against law enforcement." As they spoke, I remembered
what the Tribe had publicized over and over during the autumn of 2016 to the
media and on their website: that anyone was welcome, but no drugs, alcohol, or
weapons were allowed at the camps. Thus, their statements seemed tinged with
some ruefulness and regret that the initial protocol for the resistance to the pipeline was not followed by those who joined the camps later, and I got the sense
they felt they had lost control of what was happening during their efforts to stop
the pipeline.

The Tribe's perception of press coverage during the protests

When I asked the Tribe about their impression of how the mainstream news media
cover them—during the Dakota Access Pipeline controversy but also on a broad level—
they had many things to say. Councilman Walker stated his views clearly on the
media industry as a whole:

> I guess I could speak to that: I don't trust the media now. If there is an event
> I was there firsthand to see and experience it and the media tells something
> else that's not true? I take a look at all media outlets. You know you got
> your CNN, your Headline News, your Fox, of course they're all going
> to be biased whether left-wing, right-wing, whatever, then you got your
> Russia Today, you have your Al Jazeera, then you have your Vice News but
> all of these different news stations were here. I got to meet a lot of them
> personally, face-to-face, and one thing I will say: three months ago, I got
> interviewed by Minnesota Public Radio. They wanted to talk about dona-
> tions and GoFundMe. I talked about the spiritual aspect of the foundation
> [but it] never aired. So they all have agendas, the media.

Chairman Faith was nodding while Councilman Walker spoke, following up with the statement that "negative news... sells. I would say that there was plenty of media coverage, but more so to the negative end of it." He noted in particular Councilman Walker's earlier statement about the original intent of the Tribe to protest the pipeline in a peaceful way, then added that the media rarely covered that positive aspect:

> I would say that there were plenty of media outlets there, but to the positive end of it, very few. Unfortunately it was some of our local news media that helped with the negativeness.... Yeah they'll put you on air and they'll do this and that, but it didn't come out that way like you said. They sold what they wanted to and it was negative.... So I would say there was coverage, but... it wasn't positive. They [the media] wanted to see clashes.

In response to his statement, I asked him to confirm what I thought he was saying – that while the media tried at times to give both sides, that they did so only through the lens of conflict. To that Chairman Faith replied, "I would say that. I really would." Councilman Mauai joined in, noting on a broad level that "Councilman Walker talked about what's going to sell the news. Nobody wants to hear about a peaceful, prayerful protest that's happening here."

The idea that the news media were particularly centered on conflict and negativity sparked the following conversation about the experiences the Tribe had with local news outlets like the *Bismarck Tribune* (the subject of analysis in Chapter Four):

Councilman Yellow Fat: The Bismarck Tribune is very one-sided. Anything that bad happens down in Standing Rock, they're right here, but if we do some good things in the school...

Chairman Faith: They [recently] contacted me about four times [regarding] a police shooting down here. We can't say nothing because it's still being investigated. So they said "We called Chairman Faith four times, no answer. No response."

Councilman Yellow Fat: They kinda made the Chairman out to look like he doesn't care about it. They ask me about it, but I can't comment, because I don't know anything.

Chairman Faith: Well then we call the Bismarck Tribune [and the] Bismarck [radio station] KFYI the other day: we had probably over 300 people walk for health and wellness out here at the school. Well, guess what, they didn't respond, they didn't call me back either.... We wanted to get positive news to see all ...

Councilman Yellow Fat: We should have said it was a protest.

Chairman Faith [as good-natured laughter filled the room]: Yeah, a march or something.

Chairman Yellow Fat did make a specific statement about the emotions he feels regarding the *Bismarck Tribune* specifically, noting that he often found it to be "one-sided" with little investigative work:

> They never did get involved with interviews... interviewing people that when this pipeline breaks, it'll affect [us]? You got full focus on the arrests, you know, the protests that went into their town.... But what it came down to is [that] they all thump their U.S. Constitution, except for Native Americans or people of color.

As a result of the lack of investigation and what he felt to be the one-sided nature of the coverage, he identified the emotions he felt at the representation of the pipeline construction (as well as the resistance to it):

> I'll tell you one thing: reading the *Bismarck Tribune*, especially during the protests. Every single time I came away mad, disappointed, angry. It's just like, "do more investigation, you know, don't just put it out there." So it feeds people's anger. Of course, North Dakota is so pro-oil, they just do everything for oil and... the *Tribune* is one of a batch [of papers where] I came away feeling angry every time I read something.

Put another way, for the Tribe, the perception was clearly there that, for the local paper, oil had become more important than people – specifically the people living directly downstream from the pipeline.

The Tribe's perspective on social media during pipeline resistance

With the understanding of their perceptions regarding many of the pitfalls in traditional news, our conversations turned to social media. I asked what they perceived the role of social media to be during the events around the Dakota Access Pipeline in the fall of 2016. As with traditional journalism, the tribal members I spoke with had much to say. Chairman Faith saw both good and bad aspects of social media during the gathering, noting that it had "its pros and cons." The positives, he said, came from getting the Tribe's perspective publicized, while the negative aspect came from the spread of negative information from the camps after thousands of people started coming and lost the original message of prayer and peaceful ceremony. He also noted the fragmented and sometimes disorganized nature of social media. Ultimately, he observed, "I would say the social media... were good and bad... it may be 50/50" in terms of positives and negatives.

The Tribe seemed to appreciate that voices through social media are unedited and do not have to travel through traditional journalism before making their way to the public eye. Councilman Mauai offered that "I think in terms of social media, being utilized as a communication tool, it was a very good tool. It served our purpose." He noted that he appreciated that it was "unfiltered," but then also added "Whether or not what was being relayed… was good or bad, I think that's all in perspective to that individual. The individuals were utilizing it." When I noted that he had used the word "unfiltered," the Chairman nodded and noted that: "it was raw footage just going out there."

Considering these comments about social media being "unfiltered" and "raw," it seemed that the perception exists within the tribal members I spoke to that social media can bypass traditional structures, whether they be media outlets, big business, or politics. In this vein, Councilman Yellow Fat spoke to social media's ability to challenge old media power structures and their relationship to the oil industry, contending that "I don't have any proof of this, but I can see oil companies carry a lot of clout within the political structure. You know and so, maybe due to some pressure to keep it down… and press can only keep things down for so long." When I asked specifically about Amy Goodman's video from *Democracy Now!* (that had gone viral on Facebook), it sparked the following conversation:

> *Councilman Mauai:* I think it was great, everything that she had to present. And actually helped magnify what was happening here, because shortly after… [*looking at Councilman Yellow Fat*] was it the guy from CNN who did a couple pieces with you…?
>
> *Councilman Yellow Fat:* MSNBC
>
> *Chairman Faith [nodding]:* Yeah, Lawrence O'Donnell. I agree with that. It was beneficial.
>
> *Councilman Mauai:* [He] had done a show. What and why was happening here. So I think it was a very positive thing.
>
> *Ellen:* Do you think O'Donnell would have come here if not for Amy Goodman's video?
>
> *Councilman Yellow Fat:* It's hard to say. I mean before Amy Goodman's video, the protests at Standing Rock was just like a little excerpt. Page nine in *The Huron*. Any newspaper is like "oh, by the way there's a protest going on in North Dakota." And Amy Goodman comes out with what she did and then mainstream media started picking it up. And then other people found out what was going on so they flocked here.

This conversation was interesting because the Tribe made clear that social media can not only bypass traditional news but also work *in tandem with it*, even spurring traditional media outlets to cover news they normally would not. In addition, Councilman Yellow Fat's recognition that the placement of stories – namely, that stories about them would be placed on "page nine" – struck a chord with me, for my research on news media coverage of #NODAPL (Chapter Four) did reveal that positive stories about the Tribe were often shunted to the back pages of the local paper while negative stories often made it to the front page.

What I learned about their views on the media is, first, that context is important to them. They were keenly interested to see their words and actions placed firmly in the context of history, of their belief systems, and in terms of the treaties they held with the U.S. government. The Tribe's desire for context extended to my work as well: Chairman Faith had only one request about my representation of the Tribe in my research, and that was to not take any of their statements "out of context." In addition, the Tribe seemed very attuned to a trend noted by many scholarly works (including this one), which is that the news media lean heavily towards reporting negative news that magnifies any existing conflict. The result of this, in their view, skews the coverage away from environmental and social justice concerns and towards simply, as the Chairman put it, "selling violence." Finally, it was clear that while they perceived some detriments to social media, they also identified much that was good about it, especially in terms of encouraging mainstream media to pay attention and report on the issues that matter to them.

How they want to be portrayed by the media

Since the Tribe identified some concerns with the way in which they were portrayed – in terms of quality as well as quantity – I asked them how they would like the media to portray them. Councilman Yellow Fat led this discussion with his perceptive observation that many people in this country think about Native Americans as:

> "okay, yeah, they wear war bonnets and live in tepees." We're modern people who've held onto our past, but we don't *dwell* in our past. I mean we could be walking around here mad all the time at the government, state of North Dakota, state of South Dakota. But like what the Chairman said before... we're not putting up walls, you know we're building bridges.... So basically *we're not antiques that you put on the wall and store away.* Native Americans, we're here...from sea to shining sea [that] was once Indian land. So I think well, we need to show the people, especially media, that we've got doctors, we've got lawyers.... And what we still need is a place at the table... to negotiate with you know, water treaties, and we need a place at that table that at the United States government.

In his statement, Councilman Yellow Fat directly confronts many of the stereotypes held about Native Americans – namely, that they live in the past, that they are already extinct, that they are innately violent or prone to conflict, and that there is very little modern about them. These types of stereotypes are called out by Gibbon (2003) in his book *The Sioux: The Dakota and Lakota Nations* when he writes that many authors creating contemporary pieces about the Sioux depict them as "exotica" or as "faded reminders, historical relics, of what they once were" (p. 12).

Another set of stereotypes that they wanted to counter coming from the media were interconnected. The first harmful stereotype is that all Indigenous tribes are the same, while the other related to the perception that all Native

American nations have become rich from tribal gaming. Councilman Yellow Fat wanted it known that "not all native tribes can be clumped together." He continued by providing an example: there is a perception, he noted, held by many that because *some* Tribes have done quite well financially from gaming, *all* tribes have prospered. Chairman Faith observed that this was called "metropolitan theory" (others in the room nodded as he said this): namely, the fact that tribal casinos located in large metropolitan areas have done fairly well because they live near large, densely populated areas. Councilman Yellow Fat agreed and went on:

> Big metropolitan areas, they do very well with that. But so when somebody sees "oh man, look at these people are gaining um, $40,000 a month from, apiece from tribal gaming." Well that's not Oglala Pine Ridge, that's not Standing Rock. That's not Cheyenne River, you know. Our tribal gaming barely covers the programs that we need in place.

I got the sense from this conversation that the Tribe felt pained by the misperception that they were wealthy because it made the circumstances of every Tribe look the same, negating the need for health care or other needed services to help their members get what they need.

The Tribe also wanted to be recognized as sovereign by the media. Councilman Mauai had a clear stance on this:

> I think to circle back to your question, of how we like to be portrayed: trying to summarize it, is that we are a sovereign nation who continues to fight for the rights of our people as it is guaranteed under the existing treaties, because that's what the outside doesn't know. And it's funny because, you know you look at all of our fights: it's domestic, it's all internal, it's here in the United States. Whereas you look at other Indigenous nations on the outside, there's some type of an international influence within their governments. So the United Nations, for example... can give a special report on how we are being treated here in the United States, but the US government won't take that seriously. They will go into self-defense mode when, but this is our issue. This is something that we'll deal with right here. But under the treaties they shouldn't have. *We're sovereign, and we need to be acknowledged as that.*

Continuing with the idea that the Tribe should be recognized as sovereign, Councilman Yellow Fat noted that if the misperception were changed, it might mean "true nation-to-nation consultation" that is sorely needed.

Speaking directly to media coverage of Standing Rock in 2016, Councilman Yellow Fat opined that: "What I would like to see from the media is they need to step up and get our side too. Because you know, too many times, especially during the Dakota Access Pipeline protest, it was really one-sided or one side, you know, pro oil...." He was careful to add after this statement that "I'm not saying I'm against oil, [but simply] to be responsible and...watch over environment first."

Returning to earlier critiques of the media not covering peaceful events by Indigenous people very often, the Tribe was especially keen for me to note that the media had never publicized the ride on horseback to honor the 38 Sioux who were hanged by Lincoln in 1862. An article by Minnesota Public radio notes that this was "the largest mass execution in U.S. history... when 38 Dakota warriors were hanged from a single scaffold in Mankato. The shock waves of that mass execution still reverberate today among the Dakota people. A new documentary film remembers the 38, and also a "group of Dakota who ride on horseback each year at this time to Mankato to commemorate the executions of Dec. 26, 1862" (Steil, 2010). The hanging was the result of the "Sioux Uprising of 1862," which Erdoes (1982) attributes to the fact that, they were placed on reservations and away from their traditional hunting grounds, noting that "The Sioux were starving..." (p. 9).

The *Dakota 38* documentary was produced by Smooth Feather, which includes a message: "this film was created in line with Native healing practices. In honoring this ceremony, we are screening and distributing 'Dakota 38' as a gift rather than for sale" (the film can be found online on the Smooth Feather site (http://www.smoothfeather.com/dakota38/). While the Tribe noted that this would be great to include in the book (and thus I include it here), it also is the case that it is a documentary not easily forgotten, one that tells a rich story weaving the past with the present and intertwining tragedy with hope.

Media's lack of coverage of military service

Related to the idea that the media do not cover the positive characteristics of the Tribe, the group discussion turned to the topic of war – specifically, to the service provided by the Tribe and other Indigenous groups. Councilman Yellow Fat identified why there have always been Native American people willing to serve in the U.S. military:

> We've always found it our responsibility to protect this land against enemies. Whether it was at first The United States, Europeans that came over, or...Germany, Japan, and you know, this war we have against people attacking the environment. So, we've always been caretakers of this land: we have always found it our duty and responsibility to protect it.

In this, he found a particular irony: "we have World War I veterans who fought for this country even before they were citizens of this country."[12]

Responding to the comment about military service by Native Americans, Chairman Faith told me that the media had not covered the fact that 88 Code Talkers[13] from World Wars I and II came from the Standing Rock Sioux Tribe.

I did not know that we even had Code Talkers in the First World War, so I asked for clarification. Chairman Faith answered simply that the media tend not to publish positive stories like these. The fact that there were Native Americans serving the U.S. Armed Forces who were speaking their own language presented another irony, one well summarized by a BBC article:

> It's an irony that probably didn't go unnoticed by Choctaw soldiers fight-
> ing in World War One. While the tribe's children were being whipped for
> speaking in their native tongue at schools back home in Oklahoma, on the
> battlefields of France the Native American language was the much-needed
> answer to a very big problem.
>
> *(Winterman, 2014)*

Or, as Councilman Yellow Fat put it, "Doesn't it seem ironic? The language that they tried to kill is the language that helped saved the world in World War I and II and Korea." When I later researched this, I found a book that speaks to their service. Titled *Sioux Code Talkers in World War Two* (2017), it is authored by Andrea Page, of Hunkpapa Lakota-German ancestry whose mother grew up on the Standing Rock Reservation. On her book cover she observes that while "Many have heard of the role of the Navajo Code Talkers, but less well known are the Sioux Code Talkers, who used the Lakota, Dakota, and Nakota dialects."

Concluding the group interview

At some point during the group interview, Phyllis Young walked in. By then, Chairman Faith and Councilman Mauai had left for another meeting. Councilman Yellow Fat engaged in conversation for another few minutes and then he, too, departed, leaving Phyllis and I to talk alone. Before I describe my interview with her, a few words are appropriate to note the interaction between the various tribal members. Perhaps most notable was that the shared discussion of any topic was marked by a quiet respect and familiarity. If one tribal mem-ber interrupted another (a rare event, especially for a controversial topic with deep significance and emotion like the pipeline), it was only to affirm what the speaker had said. In addition, the tribal members constantly referred back to what the other people around the table had said as a way to support and build on other speakers' previous statements – indicating that they were listening very carefully to what other tribal members said. This is not to say that they all felt the same about every issue, but simply to note that a mutual respect seemed to exist between members, a feeling that permeated the interview session.

Interview with Phyllis Young

As we began the interview I asked Phyllis Young what she wanted people to know about her aside from her work with the American Indian Movement (and specifically with the Women of All Red Nations), her activism, and her service on the tribal council. She stated that she wanted the readers of this book to know:

> that I was a good American for 40 years and that I'm a product of my neighborhood, which is military. I was raised on [Veteran's Administration] pension and my grandfather was a Code Talker in the First World War, a Purple Heart recipient, and I have four to six uncles and fathers who were in the second World War – all Purple Heart recipients. I am a product of who we are at Standing Rock. We're more American than apple pie. And we were good Americans. And we know the [U.S.] Constitution better than most Americans.

She noted how many Native Americans have served in the military, and made clear that she feels patriotism and "and love of this country [comes] from our homeland." When I questioned her about her past-tense reference to being a "good American," she responded that she now is "forced to be a good member of the Oceti Sakowin commune, which are the Seven Council Fires, and which represent the ancestral and natural world before the written law." In particular she liked that I was from the Pacific Northwest, because she saw that many Indigenous Tribes were "trailblazers" when it comes to protecting treaty rights and using them effectively in court. She discussed what she termed "the Northwest ordinance that predates the Constitution" that was so fundamental to protecting Indigenous rights.

She continued to explain more about her changed worldview, including historical injustices like U.S. laws that forbade the speaking of native language and practicing of spiritual and cultural ceremonies including (although she did not mention it by name specifically) the "Indian Offenses Act."[14] She then said:

> The first days of my life, I was angry. I was an angry woman because of social unrest within me. And I really didn't recognize it and acknowledge it 'til I was an adult. I was one of 197 families whose homes were inundated by the building of the dams. And to this day, I have not received compensation for my grandfather's home, who was a Code Talker.

The series of dams to which she referred was part of a government plan that created a series of dams along the Missouri River, including the Oahe Dam close to the Standing Rock reservation. Phyllis knew all the details about what is known as the Pick-Sloan Plan, which, among other things, was conceived of as a way to create flood control on the tumultuous and unpredictable Missouri River.[15] Phyllis noted that, prior to the dam, she and her family were connected to the land, and that they were well fed from the arable, fertile land: "We were never hungry. We were healthy." But when the Oahe Dam was built in 1944, everything changed. Phyllis recalled the impact it had on her family and her Tribe:

They took 56,000 acres of our bottom lands and we – Standing Rock – had the largest forest of any band. The economy was critical to our grand-parents. My grandfather owned the saw mill, my mother's father owned the sawmill... 20 miles from here. Very well-to-do, very productive from that. And to go from that freedom and that pride, that dignity, being who you are, having served the United States, this country, [and] going to homelessness and all the social pathologies that were created by taking everything from our people. And the most damaging ugly part of it all, is the creation of a welfare state for our people.

Phyllis' statement echoes what Michael Lawson (2009, p. 47), in *Dammed Indians Revisited*, notes regarding the devastating consequences of the dam, which

destroyed more Indian land than any other single public works project in the United States. The Standing Rock and Cheyenne River Sioux lost a total of 169,889 acres to this project, including their most valuable rangeland, most of their gardens and cultivated farm tracts, and nearly all of their timber, wild fruit, and wildlife resources. The inundation of more than 105,000 acres of choice grazing land were affected...60 percent of those at Standing Rock.

Echoing what Phyllis had stated about the creation of a "welfare state," Lawson further explains that

damage caused by the Pick-Sloan projects touched every aspect of Sioux life. Abruptly the tribes were transformed from a subsistence to a cash econ-omy and forced to develop news ways of making a living. The uprooting of longstanding Indian communities disrupted and disorganized the social, economic, political, and religious life of well-integrated tribal groups.

(p. 51)[16]

As a result of the dam construction and other significant events in her life and the Tribe's history, Phyllis wanted me to know that "So now I am Lakota Dakota first, for the remainder of my life for my children and my grandchildren." These statements from Phyllis early in the interview represented a thread that ran through our discussion: that her identity, and the Dakota Access Pipeline, should be seen through the lens of treaty rights, historical events (including past govern-ment legislation and abuses), and court cases.

The Supreme Court, the Doctrine of Discovery, and the Dakota Access Pipeline

Phyllis placed a great deal of emphasis on the need to take many of the issues deemed important by the Tribe to the official avenue of the courts and fight there. In order to be effective in court, she contended, it was first important to understand history.

As it became clear during our conversation, she made direct connections between the decision to place the Dakota Access Pipeline upstream from the Tribe's water supply in 2016 to historical events in Europe and the U.S. centuries ago.

Phyllis began by referencing the "papal bull"[17] that formed what is now known as the "Doctrine of Discovery."[18] What she was referring to was the 1493 declaration by Pope Alexander VI, commonly identified as the *Inter Caetera*, which set the stage for the adoption of many U.S. policies and court decisions in regard to Indigenous peoples (Wolfchild, 2015; Miller et al., 2010). Pineda (2017, p. 823) briefly defines the doctrine as "the European legal concept that justified the dispossession of native lands by Europeans." Of particular note in the original bull was the papal directive noting that, if no "Christian king or prince" was already holding the lands "discovered" by European explorers, that the explorers had "full and free power, authority, and jurisdiction of every kind" over the land and the Indigenous people who lived there.[19] Later, the 1823 U.S. Supreme Court decision *Johnson v. M'Intosh* was based upon this doctrine and provided precedent for the treatment of Indigenous peoples for centuries to come.[20] In particular Chief Justice John Marshall cited the "superior genius of Europe" to justify the continued practice of subordinating Indigenous land and resource rights in the name of Christianity – a thread of logic that can be seen in modern-day court decisions in the U.S. and Canada (Wolfchild, 2015).[21] Ms Young stated that, early in her life and the lives of other Tribal members,

> We didn't know about the Doctrine of Discovery. We didn't know about the papal bulls, we didn't know the case laws that were imposed on different Indigenous nations in this country and we have all of that now. We defied the *Doctrine of Discovery* and there were 300 ministries that came here. They burned that doctrine, at the camp. So we have exercised actions that were based on the learning and the education of how the system was used against us. And so it's never going to be the same.… We are initiating lawsuits, major against the four pipelines that have um, come through here on an illegal right-of-way.

What she referenced was the burning of the papal bull at the Oceti Sakowin camp at the height of the pipeline protests in early November 2016. Pineda (2017) describes the scene:

> Episcopalian Reverend John Floberg, who was acting at the invitation of the Standing Rock Sioux, held a copy of the Doctrine of Discovery (presumably the Papal Bull titled "Inter Caetera" issued by Pope Alexander VI on May 4, 1493) and invited a committee of native elders to, if so moved, authorize the burning of the document. After a series of speeches and rituals, the document was burned to the applause and cheering of hundreds of native and nonnative water protectors…and about 500 clergy members.
>
> *(p. 823)*

Pineda goes on to note that after Pope Francis was elected, many Catholic organizations have urged the Catholic Church to

> renounce the Papal Bulls on which the Doctrine of Discovery is based. They form part of a growing movement inside and outside the Catholic Church to change what critics see as its historic apologist stance on the conquest of the Americas.
>
> *(p. 825)*

According to Phyllis, the placement of the Dakota Access Pipeline upstream from Standing Rock's water supply can be understood through the logic of colonization and the subordination of the rights of Indigenous nations. Returning to the six principles described by Miller and Riding (2011, p. 3) earlier in this chapter, it is clear that the third – "European claims to lands belonging to others by virtue of discovery are rooted in racially-based assumptions and articulated in a language that characterizes Indians as inferior savages who lack fundamental rights accorded 'civilized peoples'" – stands in direct challenge to the Doctrine of Discovery as well as the U.S. Supreme Court decision on which it is based. From this discussion, Phyllis moved directly to the idea of fighting effective court battles.

The need to uphold treaties and protect tribal rights through legal measures

Phyllis told me that in June of 2018 she would be speaking at the *International Indian Treaty Council* about treaty rights and challenges to the Doctrine of Discovery. Like Chairman Faith, Phyllis believes in taking fights to the court system and having important decisions be decided there in direct challenges to the *Johnson v. M'Intosh* decision. Also similar to the other members of the Tribe with whom I spoke, she believes in supporting Indigenous rights by honoring treaties.

According to Erdoes (1982, p. 12), historically, treaty negotiations between the U.S. government and Native American tribes have been difficult: "Always in these agreements the Indians gave up another part of their hunting grounds for a promise to remain undisturbed in the land remaining to them. Always the treaties were broken before the ink was dry." Vine Deloria (1988) states it similarly: "Indian people have become extremely wary of promises made by the federal government. The past has shown them that even the most innocent-looking proposal is often fraught with implications, the sum total of which is loss of land" (p. 49).

As we moved through our conversation, Phyllis discussed treaty rights and contemporary laws and regulations, including statutes of limitation and rights of way. She began by noting that she had been an advocate most of her life, acting as a plaintiff against Union Carbide, which wanted to mine for uranium in the Black Hills, noting that "we prevailed in that lawsuit and stopped Union Carbide from coming in. We stopped the ETSI pipeline, which was coal, slurry and water in South Dakota in 1980." As a result, she said, "we spend our lives challenging" corporations and the

U.S. government, going as far back as trying to protect the "sacred Black Hills." She described this as a "life struggle," one for which the Dakota Access Pipeline was simply the latest serious challenge to the Tribe's inherent rights – but likely not the last.

Phyllis noted the importance of the conflicts of interest she saw between the government and the Dakota Access Pipeline, observing that "there's no less than four Army Corps personnel who run DAPL." Even here she saw the historical pattern of working for the government and then working for the oil industry, citing former Vice-President Cheney's oil field ownership and his later work in the government. For her, then, "the conflict of interest is the most egregious action… because they took the information as government agents and they used it for their own personal benefit in the private sector. So we challenge this… at Standing Rock."

Speaking to treaty rights, Phyllis was especially aware of the increased solidarity between Tribes that she witnessed during the Dakota Access Pipeline struggle. She noted that "critical case… information was given to us by the chairman of the Yakama Nation, who came here," identifying specifically that she appreciated Tribes from the Northwest who "offered assistance, lawyers pro bono on treaty rights and I think that's the greatest gift that we have now and [for] the future to challenge" issues like the pipeline. I asked if she felt that the courts were more receptive in 2018 than they might have been 50 years ago, and her answer surprised me: "We're not looking for a victory: we know defeat too well." While she recognized this, she also was adamant that the Tribe would not accept any more "unilateral action" because of its treaty rights. Ultimately, she felt that "I have the obligation to create a footprint for my grandchildren to have them look and say, 'Well, my grandma did this 20 years ago. This is what she did.'"

"American are the new Indians today": the Dakota Access Pipeline, the United States government, and eminent domain

As we discussed the Dakota Access Pipeline in the broader context of protecting rights through court battles, the conversation turned briefly to the power of the U.S. government through the practice of *eminent domain*. Eminent domain is, roughly stated, the exercise of government power to appropriate private land for government or public use. Considering that the Dakota Access Pipeline is one in only a complex latticework of pipelines that crisscross the U.S. due to the heavy production of oil in the country, I brought up the young mother I had met on the plane. As noted at the start of this chapter, this woman was terrified that fracking might come to her county, despite the efforts of the people living there to fight it. Phyllis' answer surprised me: "It's critical, not only to us, but to Indian country and to America, because *Americans are the new Indians today. They are now subjected to what we have suffered, since they came here.*" She explained: "I say that because they suffer eminent domain when America wants it and you don't have a leg to stand on as an American." She spoke then of the "shock and amazement" of Americans in every state at the prospect of having their rights taken due to eminent domain and the oil industry development. For Phyllis, and the Standing Rock Sioux

Tribe, the assault on rights from the U.S. government and corporations is neither new nor a surprise, an idea returned to in the conclusion of this chapter.

Media coverage of Dakota Access Pipeline: from old to new media

Midway through the interview, before I had a chance to ask about it, Phyllis brought up "the video with the dogs," a reference to Amy Goodman's *Democracy Now!* viral video. Some members of the Tribe had been in Europe meeting with Deutsche Bank and Credit Suisse, urging executives to divest from the Dakota Access Pipeline. She observed that "Switzerland was real uppity about the whole issue. They couldn't care less, but when they saw the video with the dogs, they invited the Standing Rock delegation back." When I asked how she felt about the video she echoed what the Chairman and other tribal members mentioned earlier – namely, that she thought the video was "fantastic" because it not only got the public's attention but also gained the favorable interest of major banks. She continued by saying that journalism Goodman was subject "to the white police mentality that is still alive and well in North Dakota."

She then shifted to other forms of media, pointing (similar to other tribal members) to Lawrence O'Donnell's coverage of Standing Rock on MSNBC:

> I took a picture with him. I woke up and he was talking about treaty rights and how we own this land and how people are coming in and trespassing and yet, they're calling us the trespassers.... My husband's TV was on and I was [like], "Am I hearing things?" I had to play that over and over 'cause I was like, "I'm in 'La La Land' somewhere because I've never heard an American telecast like this".... It just reassured me that we were doing the right thing.

She saw his TV news report, which discussed treaty rights and underscored the pipeline resistance as being an issue worthy of exploration, as being so different from news coverage in the past with the Wounded Knee event in 1973 and even up to the current day:

> I think that's what made O'Donnell such a glaring experience because there was a blackout. How come nobody showed DAPL or [the use of force] on all the media in the country? Because it's owned by the executives who own the banks and who have the money for investments, big time.

I asked her about the *Bismarck Tribune* coverage of the events at Standing Rock. Her reply left nothing to be imagined: "We banished them from the camp. We kicked them out." When I asked why, she said she was uncomfortable with the paper's portrayal of Indigenous people: "we were still the Indians, the 'Cowboys and Indians' movies in their time." Continuing, she contended that:

> They [the media] are used to us being complacent and passive and not really getting up to protest, but when we do protest, we did protest.

> They're accustomed to keeping us absent in media, movies… media across the board. We're still lost in time and everyone is depicted as… riding horses and living in a teepee. So, they've locked us into that time factor and they've let us be absent.…So, we had no choice. We still have no choice. The editorials are, are very biased.

When I asked about the negative confrontation she had mentioned with the local news, she noted that there were camp rules to not take photos of anyone out of respect. Although their rules were clear and explicit, she noted that they "broke our rules," which she found disturbing and disrespectful. As a result, she would not let them stay. We moved from this recognition of the limitations of local news media to have the following exchange about new media and the potential for democratic change:

> *Phyllis Young:* I love digital democracy. I use that word all the time.
> *Ellen:* "Digital democracy"?
> *Phyllis Young [with emphasis]:* Yes.
> *Ellen:* What does that mean to you?
> *Phyllis Young:* That means that it's not my word against your word anymore. It's "seeing is believing." And we may be amateurs but.…

At this she gently joked that she didn't want a fancy modern phone because she is "an old-fashioned Indian living in a tepee so you can come find me," but that people had given her the newest smartphones and she was learning to navigate Facebook. She noted, smiling, that: "my 5-year old grandson helps me." Our exchange revealed Phyllis' view that social media can circumvent traditional news structures and tell the truth in a way that old media might not.

Conclusion of the Phyllis Young interview

Because Phyllis and I had discussed the importance of "digital democracy" when it came to Indigenous representations in the media, I couldn't help asking about a new media development that I had been discussing with my students at the University of Washington Tacoma. With new technology created by programs like the Adobe Project and the University of Washington's Graphics and Imaging Laboratory (GRAIL), people will soon be able to take a voice or video recording of someone and "puppetize" them.[22] To "puppetize" basically means to take their face and/or their voice and manipulate them to make them say and do what the owner of the software wants. New technologies like these are considered intensely controversial and are raising ethical concerns concerning misrepresentation in the era of fake news. As such, audio and visual manipulation holds great significance for anyone who enters the public stage, including those like the Standing Rock Sioux Tribe. When I asked Phyllis about this – that is, did this change her mostly positive perspective on "digital democracy"? – I noticed that she didn't skip a beat. She looked at me confidently and stated that she wasn't worried at all: "They'd never figure out how to digitize our language."

Conclusion of the Standing Rock Sioux Tribe interviews

For the Standing Rock Sioux Tribal members I interviewed, the Dakota Access Pipeline is one in a long line of injustices and privations created by the U.S. government and by corporations. Along these lines, they connected many historical events to the pipeline: the 15th century Doctrine of Discovery (and the 19th century Supreme Court case based upon that), the loss of their sacred Black Hills, the Battle of the Greasy Grass (the Little BigHorn), the largest mass execution in U.S. history of 38 Dakota by Lincoln, and the building of the Oahe Dam as part of the Pick-Sloan Plan. The Tribe also saw a relationship between the pipeline and events that may occur in the future, including the Trump administration's suggestion that Tribes should be reclassified as another "race" instead of as a sovereign nation. For the Sioux in particular, this would have far-reaching negative consequences, including crucial access to health care and losing the ability of self-governance. For many of them, the significant inter-tribal solidarity gained from the pipeline fight will help them to resist any of Trump's new plans that involve them.

Seen in this light, the Tribe moved seamlessly and fluidly between the present, past, and future as a way to understand the Dakota Access Pipeline. It is not to say that they dwell on their past, but it is clear that for them it is all tied together: there is no separating one from the other, and the placement of the pipeline directly upstream from their water is an event that cannot be divorced from the context of who they are as a people and what has happened in the past. Perhaps one way to understand this perspective returns us back to Phyllis' discussion of the Doctrine of Discovery, for it created a centuries-long predicament for Indigenous groups that ultimately "perpetuates a second-class national status for tribal nations and relegates individual Indians to a second-class citizenship status, because it strips tribes and Individuals of their complete property rights" (Wilkins and Lomawaina, 2001, p. 20). The placement of a pipeline on their ancestral lands directly upstream from their sole source of drinking water could well be seen as an instantiation of always being treated by corporations and the U.S. government as "second-class citizens," and it seemed the Tribe was well aware of this.

The Tribe is also very focused on treaty rights and fighting future battles, if they must be fought, in the courts. Deloria (2011) noted that the treaties reflected an agreement or contract between sovereign nations, one that was meant to be immutable unless both parties agreed to dissolve it:

> the Indians in fact had these particular rights, that the United States orally assured them that they had these rights, that a particular section of the treaty guaranteed these rights, and that the United States, and often its

courts, will attempt to blur the issue so that the rights are neglected, for-feited, abandoned, or deliberately sidestepped whenever Indians ask that they be enforced.

(p. 75)

Despite the past history of broken treaties, the Tribe considers these govern-ment-to-government contracts as an important part of their future in holding on to their rights.

Finally, the Tribe had considered the role of media in all of this. They seemed to uniformly appreciate Amy Goodman's *Democracy Now!* video on social media as well as Lawrence O'Donnell's searingly critical coverage on MSNBC of the U.S. government's violation of the Tribe's rights to treaty consideration as well as their human rights to clean water. While they criticized local news, and particu-larly the *Bismarck Tribune*, theirs was a nuanced perspective, one that understood the contours of power and influence when it comes to the interaction between oil, the state, the media, and the (mis)representation of Indigenous people.

Interview with journalist and author Mark Trahant

Independent journalist, editor, scholar, and "Twitter poet" Mark Trahant emerged as a clear choice to interview for this research for many reasons. First, he is one of the most experienced journalists who has written extensively on many different Indigenous issues, including health care reform, sexual abuse by priests in Native Alaskan villages, and politics. He was the editor for the Seattle Post-Intelligencer and a journalist at numerous news outlets, including the *Navajo Times*, *The Salt Lake Tribune*, the *Sho-Ban News*, and the *Seattle Times*. He also served at the helm of the *Native American Journalists Association* (NAJA), which recognized him with the Richard LaCourse Award for his "Native elec-tions coverage, and dedication to NAJA as a lifetime member and leader."[23] He is the author of the blog "Trahant Reports" as well as many scholarly publications, including *The Story of Indian Health is Complicated by History, Shortages & Bouts of Excellence* (2018), *The Last Great Battle of the Indian Wars* (2010), *The Constitution as Metaphor* (2000), and the book chapter "A Tribe with a View" (2000).

While working at the *Seattle Post-Intelligencer*, Trahant was responsible for one of the first "viral" videos (this was before the YouTube era) in 2004 by asking then-President George Bush a simple question: "What do you think tribal sover-eignty means in the 21st century, and how do we resolve conflicts between tribes and the federal and state governments?" Bush, famously, was unable to answer the question sufficiently, and in the process revealed much about power imbal-ance between the U.S. government and Native American Tribes.

Being a member of Idaho's Shoshone-Bannock Tribe, he also brings an impor-tant Indigenous perspective to the Standing Rock protests. While he was still the Charles R. Johnson Endowed Professor of Journalism at the University of North Dakota, he proposed to hold a conference to discuss the significance of social media for the events at Standing Rock. The university appeared to stall his request due to concerns over the political nature of it, and as a result Trahant said he would quit. The university capitulated and Trahant held his conference, titled "Standing Rock & the Media" in April 2018, bringing together students, journalists, scholars, and activists. At the end of the academic year, Trahant chose to leave to become the editor at *Indian Country Today*, which is where he is as of this writing.

I sat down with Mark Trahant in his office at the Department of Communication on the University of North Dakota campus in the afternoon of April 26. Although he could have stayed behind his large desk by the window, he joined me in a nearby chair. As an experienced journalist who has done hundreds (if not thousands) of interviews, he was much more at ease with technology than me: he suggested using my phone (with extra battery pack) instead of the digital recorder I had, making what I originally felt to be the latest technology seem more like a dinosaur. During our interview, students would pop in to say "hello" and ask about their final projects in an easy flow. I was grateful for his time, since he was busy working three jobs (including teaching but also having just taken the helm at *Indian Country Today*). I told him about the findings of my research on the *Bismarck Tribune*, *New York Times*, and *Indian Country Media Network* (presented in Chapter Three). Then I asked him some questions and he shared his thoughts on pipelines and the influence of Big Oil, media coverage of the Standing Rock resistance, the politics of interviewing U.S. presidents, and the power of Indigenous–produced media.

The Dakota Access Pipeline

Mark indicated that he had researched the environmental ramifications of oil trains long before the controversy erupted over the placement of the Dakota Access Pipeline upstream of the Standing Rock Sioux Tribe's water supply, and that he had changed his opinions throughout his research:

> I started writing about it before there was a protest.... For me it really came out of the controversy with the XL pipeline[24] and Keystone XL. And I actually had to reverse course because initially I was worried about just the massive number of both coal and oil trains in the west, and the first piece I wrote was on that perspective. As I started researching, I was really swayed by the UN climate change report that talked about permanent infrastructure. The UN report was saying if we're going make a change on climate we've gotta stop doing permanent things. And so then I kind of switched, and then...I did several pieces on Keystone, and then as part of that I started to do the Dakota Access pipeline.

Here, Mark made clear that his interest in writing on pipelines stemmed from his initial perspective that pipelines were safer due to the danger that trains pose (in an immediate sense) to the environment and local communities. Later, however, he attributed his pivot to his understanding that the creation of permanent infrastructure means that we become locked into a relationship with fossil fuels.

Media coverage of the pipeline resistance at Standing Rock: local news media

As a journalist with four decades of experience, Mark had much to say about how traditional news media covered the events at Standing Rock. He first discussed the weakened and shrunken state of the newsroom of local newspapers (a trend noted in the opening chapter of this book), especially when it came to the *Bismarck Tribune*. In particular, he observed that "the shrinkage they have experienced meant that they didn't have many reporters to be there." In response to my query about the both quantity and quality of the coverage by local papers like the *Tribune*, he observed that:

> The quantity was interesting because the press really started locally with the arrest of [Tribal Chairman] Dave Archambault. And that was pretty much it. Local press was really caught up in the media shrinkage where they had days where they didn't have anyone in the news room.... Had this been Wounded Knee[25] they would have sent three reporters, they would have been there the whole time, and they would have been filing [stories] every day. But the Bismarck community didn't have three reporters. I mean it just didn't have the resources.... When [*Bismarck Tribune* reporter] Grueskin wasn't available, they had no one to cover the issue... and when she had a day off they didn't have someone to back her up.

Here, Mark identified an important trend – a growing lack of resources for local newsrooms – that was a significant contributor to how much local media like the *Bismarck Tribune* covered the Standing Rock resistance to the pipeline. While he characterized the shrinkage of media newsrooms and staff as "extraordinary, particularly for local news," he observed that international news outlets, which were "actually the first there in a big way," did not seem to be similarly constrained.

When I asked about a perception by one Standing Rock tribal member that the media seemed to be in "lockstep" with the state and with the oil industry, Mark replied that while "in certain areas they were in lockstep," what seemed more true to him was that that they did not have editorials that would provide much-needed perspective, they did not have "native voices" in op eds, and they were constrained by both lack of resources and inexperience. He stopped short of saying that the local news media had become "confrontational," but said he recognized a key difference in the relationship reporters had with Native American protesters in the past: "If you look back at the coverage of Alcatraz,[26] for example, the press completely identified with Alcatraz. I remember a story where a reporter for KGO TV loaded up groceries to go to the island, because [the Native American protesters] had run out."

"Alcatraz" refers to when Indigenous activist Richard Oakes and a group of supporters claimed the island off San Francisco for Native American people. It then turned into a larger occupation that lasted many months until June 1971 (Johnson, 2015). In referencing the protests there, Mark identifies a significant shift from reporters empathizing with the social justice concerns of those they covered to a more detached, if not confrontational, approach. Coming back around to the results of my research on how the local news media treated the protests at Standing Rock, he did note that he thought their framing of the story was "weak" and that he thought the frames I identified in the *Bismarck Tribune* in particular were "pretty accurate."

I asked Mark a question regarding how much he thought the oil industry had shaped local media coverage of the Standing Rock resistance. While he didn't feel that he could make explicit ties between the news media and the oil industry, Mark did note the more structured relationship the entire state of North Dakota has with oil, noting that "The home influence is huge." Wanting clarification, I asked him to elaborate. His response was that the oil industry is

> such a huge employer and tax provider. What's interesting about that is this is a state that gives oil what it wants, so it doesn't really pay the freight for all of this. It's not taxed as heavily as say, Alaska.

When I asked him what he thought that meant, his response was simple: "The state does what oil wants it to do."

Media coverage of the pipeline resistance at Standing Rock: national news media

After we discussed local news, our conversation turned to national media, with Mark noting the absence of many major news outlets at the height of the protests:

> One of the press stories that struck me was that the *New York Times* called here, not to talk to me, they talked to a colleague, and wanted to get a stringer[27] there for the whole time. But they were looking here and not Standing Rock. They weren't looking to the Native American press saying, "Here's someone we could pick."

In so stating, Mark highlighted the omission of the Native American voice by the paper, even though inclusion of Indigenous press members could have provided needed context and insight on the events that unfolded.

He also observed that the paper did not cover the treaty issue much, but believed that they mitigated this silence through their editorial board:

> I have to look this up to be sure, but I think the editorial was ahead of the newsroom in terms of that frame. I think the newsroom was catching up. So most newspapers, and certainly the *New York Times* are divided in church and state,[28] and the editorial board is church, and they don't talk with the news side. When I was editorial page editor [at the *Seattle Post Intelligencer*] I loved

scooping the newsroom. It just was great fun. And this had that element where the news side wasn't paying as much attention [to Standing Rock]. The difference is the editorial board works directly for the publisher, they have no ties to the managing editor. They're completely independent. And so when their editorial came out I think it really kind of sent an electric spark through the *New York Times'* newsroom.

Mark noted that TV news played a particularly important role, identifying Lawrence O'Donnell as being "hugely important." Specifically he noted that

He came out and... really helped frame the discourse. He did a good job. He was pretty deep on the issues he studied. He got to the aspect of treaties which was really key to, I think, the framework for the story. So yeah, I would put him on the list of folks who did it right.

As noted earlier, the Standing Rock Sioux Tribe also had singled out O'Donnell's coverage as being both surprising and appreciated in its treatment of treaty rights and its call for human rights. Thus, one clear result from this research is how much O'Donnell's words meant from the Tribe's perspective (and, in the case of Mark Trahant, from a journalist's perspective too).

New media and social protest

In all, Mark perceived two clear benefits from social media in relation to the events at Standing Rock. First, he noted that it was Facebook that got then-President Obama informed and involved with what was happening at Standing Rock in a way Mark found "stunning" as a journalist:

I mean it was clear from social media that the president of the United States, who had been to Cannon Ball, who had someone from Standing Rock who had worked for him, was not informed about the situation because he was asked by someone who read it on Facebook.[29] And the White House press corps had no clue. His answer is so stilted that it's really clear he didn't know what was going on. He got up to speed very quickly after that, but... it's funny because I can think of three chapters in history, one of which involved me, where presidents of the United States were caught blindsided by issues in the Indian country. And I think this was one of them.

Mark categorized this interaction as one of only a handful of "chapters in history...where presidents of the United States were caught blindsided by issues in Indian country." The other interaction, Mark noted, was his now-famous question to then-President George Bush about tribal sovereignty. Bush's answer, as YouTube later memorialized, was "Tribal sovereignty means that: it's sovereign. You're a... you've been given sovereignty and you're viewed as a...sovereign entity." Trahant's question – paired with Bush's stilted response – revealed much about the federal government's perspective on Tribal sovereignty: that it was *given* by the U.S. government, which suggested that sovereignty was not an inherent and unalienable right held by Native American tribes. When I asked Trahant about the motivation behind his question – that is, did he know it might cause such controversy? – he replied that there was no artifice or agenda: "I had no idea. I really thought he would be prepared for such a question. I wanted something that he, personally, would be able to answer."

Returning to the events that unfolded at Standing Rock, Mark, saw social media and the Internet as having "inform[ed] the President, and got[ten] the White House involved." The second benefit was, in Mark's words, "the extraordinary amount of communications by people on the ground who were solid in their factual reporting," adding that he found it significant that this reporting came from those who weren't trained journalists. Contextualizing this in history, he noted that while earlier Indigenous activism and protest was reliant on the traditional news media to cover them, Indigenous activists today can employ social media:

> What made Standing Rock different, even from *Idle No More*[30] which was just a year or two before, was Facebook Live. And Facebook Live started just three weeks before Standing Rock. And so suddenly people could go on and have a national audience [trending at the] top of Facebook at Standing Rock.

He highlighted Dallas Goldtooth from the *Indigenous Environmental Network* in particular for his effective and clear breakdown of the events of every day. In terms of gaining a large audience, Mark highlighted law enforcement's use of water cannons on the protesters: it happened "in the middle of the night, with 200,000 to 300,000 people watching it live."

In addition to the benefits, Trahant listed some disadvantages, or challenges, when it came to the protests. The first detriment was the spread of rumors on social media like Facebook without being verified. "Some of them were just outlandish...things like the sheriff spraying a potent chemical on top of us at night." When I asked Mark about one of the rumors about new media that had been flying around – namely, the idea that Facebook was censoring protesters who received a message that their posts were "denied based on content." His answer surprised me:

> I was actually, of all things, at Facebook that week. And the funny thing was we were with the news side in Facebook and they loved this story. They loved that Standing Rock was a story, and getting world attention through Facebook. There's no way to confirm this, but what I think really happened is the pipes can only take so much data, and 15,000 [people] are all on Facebook, and it crashes things. It wasn't just Facebook, it was cellphones. I mean just that many people in an area that was never designed to handle that kind of traffic.

Mark also identified some interesting challenges faced by law enforcement, which I interpreted as being a note of caution to those who would use social media to try to gain support for their side: "the use of social media by law enforcement… was almost hilarious because it was so weird." When I asked for an example, he replied that "they've scrubbed it now so you can't even see it, but Morton County would just try to tell their version of the story on Facebook… it was hateful, it was unprofessional and it was just something else."

New media, Amy Goodman's video, and civil rights

During the interview, I specifically asked about Amy Goodman's *Democracy Now!* video that publicized security company dogs biting Indigenous protesters. As noted earlier in the book, the video quickly went viral after being posted on Facebook, prompting many mainstream news outlets to end their media blackout on the issue. It was in fact that video that drew my attention to the issue, and led me down the path to this book, so I wanted to ask what Mark's perceptions were regarding the potential impact of the video. He praised Goodman, stating that "Amy was bold enough to go where other reporters, particularly TV crews, wouldn't go. And so, she had access. She was lucky in the timing to be there when all that happened." He continued by noting that the "dog story was probably a turning point in press coverage. I think more and more people then realized this is a big deal, we need to be there."

Significantly, Mark noted that some of the key symbols – police dogs, protesters, military-style law enforcement[31] – in Goodman's video evoked those images from 1960s civil rights' protests in Birmingham. In other words, the images of brutality against people of color provided a familiar, well-worn frame of reference for those watching. As Mark put it:

> the images shifted the story because anybody born after 1960 immediately knew what those images were. Bull Connor, Alabama… *It was a context that was American.* And that put it into a story frame that I don't think North Dakota could ever get out of. In fact, if I had been in North Dakota, well I would have been, "Well let's investigate this right now. Let's shut down everything until we figure this out."

On the role of Indigenous-produced media

Nearing the end of our interview, I asked Mark about his perception of the role of Indigenous-produced media and news written by Indigenous journalists. Overall, despite some prior issues with how outlets like *Indian Country Today* were financially supported (problems with click bait and the like), Mark praised Indigenous-produced journalism:

> Well *Indian Country Today* in Standing Rock is huge… they stepped up big time for Standing Rock. They had people on the ground from the

beginning. They'd take everything that was written. They really made it their mission to get the Indian voice out in a very, very big way. And to give people the respect of telling their stories. And that's probably the first time in history something like that has happened in that regard.

Somewhat ruefully, Mark noted a key distinction in the behavior of Indigenous journalists versus those at mainstream outlets in the example of the hantavirus outbreak in The Four Corners[32] in the 1990s (the hantavirus was a respiratory illness spread by a deer mouse). Mark grimaced slightly when I recalled the name given to the illness by the mainstream press due to the first people to have suffered from it – the "Navajo Flu" – but then clarified the point he was trying to make: "that was a similar [issue] where the native press was covering it. In fact it's funny, a disease that's killing Indians and the Indian reporters all run *to* it." Ultimately, Mark reiterated his point: that Indigenous-produced journalism (and especially *Indian Country Today*) covered the resistance to the Dakota Access Pipeline very well, in terms of both quantity and quality: "I can't say enough about how they did Standing Rock."

The state response to Standing Rock

As Mark and I wrapped up our time together, he made a statement that once again compared the past to where we find ourselves now – a statement about the past that could be read as advice for today. Specifically, he drew attention to how much he thought had changed in terms of the official response to Indigenous protest. Echoing his earlier statement regarding the connection between the civil rights era and the #NODAPL movement in 2016, he commented that:

> One time the Smithsonian asked me to interview people from the Nixon administration, and I interviewed Leonard Garment[33] who many people thought was the conscience of Richard Nixon. And Garment told me that… the greatest power of a government is to send in police or troops. And he was talking about the takeover of the BIA[34] at Alcatraz, specifically. And he said his argument to Richard Nixon, what carried the day, was that we need to be patient and show that we're above that. And Alcatraz went 18 months. And you think about that approach, in North Dakota had that been it, it would have been a very different story. But North Dakota was impatient, they wanted the oil right now, and so just the very nature of it was challenging.

Conclusion

The interviews with the Standing Rock Sioux Tribe and editor and journalist Mark Trahant contained many similar veins of thought. One that ran through the interviews was the power of Big Oil. As Mark noted, "this is a state that gives oil what it wants." Similarly, Councilman Yellow Fat stated his belief that "oil companies carry a lot of clout within the political structure." Remarks like these echo the point made by Juhasz in the first chapter of this book regarding the power of the *oilygarchy*: "Big oil has simple needs. It wants to explore for, produce, refine, and sell oil and gas wherever possible without restriction....It does not want to be slowed down or financially burdened by government bureaucracy, environmental laws…or concerns for human rights" (2008, pp. 209–210). That both the Tribe and Mark recognized the power of the oil industry speak to the common perception held by many that the state chose oil over the concerns of people.

The media were discussed in the interviews as a source of frustration (especially for "old" media like journalism) and hope ("new" media). While local newspapers may not have a critical element and may only cover the negative, fostering serious misrepresentations, perceptions were there that social media could help to bypass the older structures and provide an "unfiltered" truth. As such, the interview with Mark Trahant helped to contextualize the Tribe's statements about the bias they perceived in coverage. Specifically, what Mark noted was *change* from the way things used to be: lessening resources for local media and changing attitudes by journalists in the way they relate (or, perhaps, avoid relating) to the people they cover.

Finally, it would be a mistake to consider the Dakota Access Pipeline as an isolated event. What I mean by that is that it can be placed in numerous contexts as a way to understand how it was that a pipeline came to be placed directly upstream from the Tribe's drinking water source: the American addiction to oil, creating a "need" for it that overshadows the rights of people and communities; the U.S. history of minimizing or ignoring the very real needs of people of color; and a changing media landscape that severely constrains the press. The Standing Rock Sioux Tribe thus found themselves at the epicenter of a perfect storm, one involving a long history of injustice, a society soaked in oil, and a media system crippled due to its close ties to corporations and the oil industry.

Notes

1 Lindlof and Taylor (2019, p. 11) refer to the problems inherent in the scientific approach as "positivism's compromised status," which led to the rise of interpretivism.
2 Geertz (1973) himself saw this as relating to semiotics due to the numerous significations that are created and can be partly revealed and understood through the process of *interpretivism* and *thick description*. Thick description, Geertz notes, originates in philosopher Gilbert Ryle's distinction between superficial "thin" description and deeper, more contextual "thick" description.
3 Lindlof and Taylor list many more characteristics of the interpretivist approach, but these are the characteristics most germane to my research.
4 Towards this end, Guest et al. (2013) make the well-taken point that "there are many different kinds of data collection and analysis techniques, as well as the diversity of theoretical and epistemological frameworks that are associated with qualitative research" (p. 3).
5 The song was featured on a 1969 album of the same name featuring Lynn on the cover sporting braids, a leather dress with fringe, and a turquoise necklace.
6 Available at https://docs.house.gov/meetings/AP/AP06/20180510/108265/HHRG-115-AP06-Bio-MauaiB-20180510.pdf

7 According to Bressan (2017), "many Native American nations of the American plains, such as the Blackfoot, Kiowa, Mandan, and Lakota, used chronologies with pictograms of notable events to record the passing of the seasons. These events include important battles, a good hunt, or the death of a leader, but also observations of the surrounding environment." Erdoes (1982) notes that typically these documents were kept by elder men in the Tribes and that "they were a form of picture writing done on tanned buffalo hides. Each picture depicted an event having taken place in a particular year" (p. 5).

8 Much of this background on Wounded Knee that I wrote here was provided by Erdoes (1982), who provides a broad treatment of the events of the massacre.

9 As Greene (2016) recounts, the run began in early April, 2016: "weeks after the start of the Standing Rock encampment, [a Kul Wicasa Lakota named Danny] Grassrope saw a flier on Facebook for a run – 564 miles from Cannonball, NC, to Omaha, NE, over eight days – to raise awareness of the fight against the DAPL. An old friend of his, Bobbi Jean Three Legs, was organizing it with a group of Native youth called ReZpect Our Water (ReZpectOurWater.com).

10 Sitting Bull of the Hunkpapa was an elder and "statesman" who "united all the Sioux tribes…defeating Custer and the Seventh Cavalry at the Little BigHorn" per Erdoes (1982, p. 15).

11 The Battle of the Greasy Grass is more colloquially referred to as the "Battle of the Little Bighorn" or, from a different perspective, "Custer's Last Stand," where Custer's 7th Cavalry decisively lost to Native Americans (primarily of the Sioux Nation) over treaty violations in the Black Hills.

12 Native Americans became categorized as U.S. citizens in 1924 in the *Indian Citizenship Act* years after the end of World War I. This was laced with much controversy, as many Native Americans had been considered citizens before this and many had not wanted to be considered citizens despite the Act. As *Indian Country Today* (2014) has noted, within this there were more controversies. First, even after the Act was passed, "some Native Americans weren't allowed to vote because the right to vote was governed by state law." Second, the move was largely seen as a way for the federal government to "absorb Indians into mainstream American life." Finally, as David Wilkins (2014) observes in *Dismembering Natives*, "Although Native individuals had U.S. citizenship thrust upon them without their consent, they retained citizenship in their own tribal nations." Thus, the Act itself remains mired in tension.

13 Code Talkers were those Native Americans who spoke their own language, in the process making a "code" that the Germans found impossible to break. Code Talkers existed in both World War I and II.

14 As Erdoes (1982) notes, "under the 'Indian Offenses Act,' participation in Sioux religious rituals was punishable by jail" (p. 20).

15 As noted by Lawson (2009), there were three parts to this plan: creating levees that would help with flood control; constructing a series of tributary dams; and creating five dams on the Missouri itself.

16 It is important to note the significance held by Phyllis and the Tribe regarding Lawson's book. At one point in the interview she mentioned that she had placed *Dammed Indians Revisited* in the Standing Rock schools so that they would learn about the Oahe dam and its impact on their people.

17 "Papal" means of or relating to the pope, while "bulls" is derived from "bulla," or the lead ball used to anchor the papal seal (Wolfchild, 2015).

18 According to Miller et al. (2010, p. 27), "The Doctrine of Discovery was the international law under which America was explored and was the legal authority the English Crown used to establish colonies in America." Discovery allegedly passed "title" to Indian lands to the Crown and preempted sales of these lands to any other European country or individual and granted sovereign and commercial rights over Indian nations to the Crown and its colonial governments. They go on to note that "According to the Court and the Doctrine, the discovering European nation gained real property rights to native lands and sovereign powers over native peoples

and governments merely by finding lands unknown to other Europeans and planting their flag in the soil (p. 4). Finally, they observe that while "discovery was an international legal principle designed only to control the European nations... clearly, however...Indigenous peoples and nations have felt most heavily its onerous burdens" (p. 5).

19   The *Inter Caetera* was translated into English and published in full by Great Neck Publishers (2017), from which these quotes are drawn.

20   Miller et al. (2010, p. 26) provide a particularly cogent summary of how the Discovery doctrine was used in the U.S., contending that "The legal evidence of American history shows that the establishment of the 13 original English colonies and the 13 original American states and the creation and expansion of the United States thereafter was based on the Doctrine of Discovery. The American Founding Fathers were well aware of this international law and utilized it while they were part of the colonial English system. They then naturally continued to use Discovery as the law of the new United States.... American leaders used Discovery to justify their claims of property rights and political dominance over the Indian nations and citizens."

21   In particular director Wolfchild and Steven Newcombe (upon whose book Wolfchild's documentary is made) notes the controversy surrounding tribal rights on Great Turtle Island. LaDuke (2017) writes that "Tribal nations across Turtle Island have been emboldened by the resistance movement at Standing Rock, and are taking unprecedented actions to protect our lands, waters, sacred places, and treaty rights. In the Great Lakes, Native communities have been fighting for years to shut down old oil pipelines that threaten our territories and to resist Canadian energy company Enbridge's plans to expand a massive network of pipelines through the region....Every potential pipeline that would move tar sands or fracked oil to Lake Superior and the other Great Lakes would run through Ojibwe reservations or treaty land."

22   Anyone wanting to learn more about this can listen to the Radio Lab podcast "Breaking News," available here: https://www.wnycstudios.org/story/breaki ng-news

23   Information about the award is available on the NAJA website: http://www.naja. com/news/m.blog/509/naja-selects-mark-trahant-as-2018-naja-richard-lacours e-award-recipient

24   A seeming reference to the Canada XL pipeline.

25   This is a reference to the takeover of the small town of Wounded Knee in South Dakota in 1973 by Russell Means and other Oglala Sioux Lakota members to protest the breaking of treaties by the U.S. government. At that time, numerous reporters from many outlets were on the scene and wrote hundreds of stories on the issue, even if the quality was lacking context and sometimes filled with stereotypes (Christians et al., 2011). Ultimately, the Federal marshals and National Guard exchanged bullets with the activists and then cut off supplies to the town. When a sympathetic supporter tried to air drop supplies to them, federal agents opened fire on them, causing the first of several deaths in the standoff – a Cherokee man (Chertoff, 2012). Of course, Wounded Knee had its own significance prior to this event, in what is known as the Wounded Knee Massacre where hundreds of Lakota Sioux were killed by the U.S. Cavalry troops.

26   Johnson (2015) notes that there were previous attempts to reclaim the island: "In actuality, there were three separate occupations of Alcatraz Island, one on March 9, 1964, one on November 9, 1969, and the occupation which lasted nineteen months which began on the 20th of November, 1969."

27   A "stringer" is the colloquial news term for a journalist or photographer who works freelance, creating and submitting reports or media material for pay on a piece-by-piece basis.

28   Although often the phrase "church and state" refers to the separation between the advertising staff and the newsroom, Mark here uses it to refer to the strict separation between the editorial board and the news room.

29 Mark Trahant refers here to Obama's visit to Indonesia in early September, 2016 – shortly after Amy Goodman's video showing dogs biting the protesters – where a reporter from Malaysia asked for his position on what was happening at Standing Rock. Obama's response was that "I can't give you details on this particular case: I'd have to go back to my staff and find out how are we doing on this one."

30 "Idle No More" is an environmental justice movement that emphasizes the need to honor "Indigenous sovereignty and to protect the land and water" (from www.idle-nomore.ca/vision).

31 The fact that it was a private security firm and not law enforcement did not change the civil rights' frame of reference.

32 In 2000 Mark had written about the symbolism in the Four Corners, noting that the proposed interpretive center planned by the U.S. government seems to ignore that "all four of the corners are in Indian country – not on state land. The northeast corner is on Ute land; the rest under the jurisdiction of the Navajo Nation" (p. 2).

33 Leonard Garment was a Wall Street lawyer and adviser to President Nixon during Watergate.

34 The Bureau of Indian Affairs.

References

Bharadwaj, L. (2014). A framework for building research partnerships with first nations communities. *Environmental Health Insights, 8*(8), 15–25. 10.4137/EHI.S10869.

Brewer, G. (2018, April 27). Trump takes a hard line on tribal health care. *High Country News*. Retrieved from https://www.hcn.org/articles/indian-country-news-trump-takes-a-hard-line-on-tribal-health-care

Chertoff, E. (2012, October 23). Occupy wounded knee: A 71-day siege and a forgotten civil rights movement. *The Atlantic*. Retrieved from https://www.theatlantic.com/national/archive/2012/10/occupy-wounded-knee-a-71-day-siege-and-a-forgotten-civil-rights-movement/263998/

Christians, C. G., Fackler, M., Richardson, K., Kreshel, P., & Woods, R. (2011). *Media ethics: Cases and moral reasoning* (9th ed.). Boston, MA: Allyn & Bacon.

Deloria, V. (1988). *Custer died for your sins: An Indian Manifesto*. Norman, OK: University of Oklahoma Press.

Deloria, V. (2011). The United States has no jurisdiction in Sioux country. In S. Miller & J. Riding (Eds.), *Native Historians write back: Decolonizing American Indian History* (pp. 71–77). Lubbock, TX: Texas Tech University Press.

Eagle, J. (2016). Declaration of Jon Eagle, Sr. in support of motion for preliminary injunction. District of Columbia. Retrieved from https://earthjustice.org/sites/default/files/press/2016/Declaration-of-Jon-Eagle-Sr.pdf

Erdoes, R. (1982). *Native Americans, the Sioux*. New York: Sterling Pub. Co.

Galleta, C. (2013). *Mastering the semi-structured interview and beyond: From research design to analysis and publication*. New York: New York University Press.

Geertz, C. (1973). *The interpretation of cultures: Selected essays*. New York: Basic Books.

Gibbon, G. E. (2003). *The Sioux: The Dakota and Lakota nations*. Malden, MA: Blackwell Pub.

Greenwood, D., & Levin, M. (2005). Reform of the social sciences, and of universities through action research. In N. Denzin & Y. Lincoln (Eds.), *The SAGE Handbook of Qualitative Research* (3rd ed., pp. 43–64). Thousand Oaks, CA: Sage Publications.

Guest, G. (2013). In E. E. Namey & M. L. Mitchell (Eds.), *Collecting qualitative data: A field manual for applied research*. Thousand Oaks, CA: SAGE Publications.

Hagerty, Silas (dir.), in conjunction with Smooth Feather Productions. (2012). *Dakota 38*. Retrieved from https://www.youtube.com/watch?v=1pX6FBSUyQI

Indian Country Today (2014). Native History: Citizenship thrust upon Natives by U.S. Congress. *Indian Country Today.* Retrieved from https://newsmaven.io/indiancou ntrytoday/archive/native-history-citizenship-thrust-upon-natives-by-u-s-congress-diZuPnKVq0C3vugN0JlgQQ/

Johnson, T. (2015, February 27). We hold the rock. *National Park Service.* Retrieved from https://www.nps.gov/alca/learn/historyculture/we-hold-the-rock.htm

LaDuke, W. (2017, January 31). Winona LaDuke on new ways to keep pipelines out of the great lakes. *Yes! Magazine.* Retrieved from http://www.yesmagazine.org/planet/tribes-find-new-ways-to-keep-pipelines-and-their-oil-out-of-the-great-lakes-20170131

Lawson, M. L. (2009). *Dammed Indians revisited: The continuing history of the Pick-Sloan Plan and the Missouri River Sioux.* Pierre, SD: South Dakota State Historical Society Press.

Lindlof, T. R., & Taylor, B. C. (2019). *Qualitative communication research methods* (4th ed.). Thousand Oaks, CA: SAGE Publications, Inc.

Miller, S., & Riding In, J. (2011). *Native historians write back: Decolonizing American Indian history.* Lubbock, TX: Texas Tech University Press.

Miller, R. J., Ruru, J., Behrendt, L., & Lindberg, T. (2010). *Discovering indigenous lands: The doctrine of discovery in the English colonies.* Oxford: Oxford University Press. oso/9780199579815.001.0001

Nicholson, B. (2017, September 28). Tribal head who led Dakota Access Pipeline fight voted out. *Associate Press.* Retrieved from http://www.post-gazette.com/powersource/latest-oil-and-gas/2017/09/28/Tribal-head-who-led-Dakota-Access-pipeline-fight-voted-out-3/stories/201709280160

Page, A. (2017). *Sioux code talkers of World War Two.* New York: Pelican.

Pineda, B. (2017). Indigenous pan-Americanism: Contesting settler colonialism and the doctrine of discovery at the UN permanent forum on indigenous issues. *American Quarterly, 69*(4), 823–832. 10.1353/aq.2017.0068.

Rowley, J., Jones, R., Vassiliou, M., & Hanna, S. (2012). Using card-based games to enhance the value of semi-structured interviews. *International Journal of Market Research, 54*(1), 93–110. 10.2501/IJMR-54-1-093-110.

Shohat, E. & Stam, R. (2014). *Unthinking eurocentrism: Multiculturalism and the media* (2nd ed.). New York: Routledge, Taylor & Francis Group.

Sprague, D., & Sprague, R. (2015). *Standing Rock: Lakota, Dakota, Nakota nation.* Charleston, SC: Arcadia.

Steil, M. (2010, December 4). Documentary recalls killing of 38 Dakota warriors in 1862. *Minnesota Public Radio.* Retrieved from https://www.deseretnews.com/article/700094755/Documentary-recalls-killing-of-38-Dakota-warriors-in-1862.html

Wilkins, D. (2014, May 16). Dismembering Natives: The violence done by citizenship fights. *Indian Country Today.* Retrieved from https://newsmaven.io/indiancountry today/news/opinions/dismembering-natives-the-violence-done-by-citizenship-fights/

Wilkins, D., & Lomawaima, K. T. (2001). In K. T. Lomawaima (Ed.), *Uneven ground: American Indian sovereignty and federal law.* Norman, OK: University of Oklahoma Press.

Winterman, D. (2014, May 19). World War One: The original code talkers. *British Broadcasting Corporation.* Retrieved from https://www.bbc.com/news/magazine-26963624

Wolfchild, S.P. (2015). *The doctrine of discovery: Unmasking the domination code.* [Video/DVD] Morton, MN: 38 Plus 2 Productions Distributor.

8

DID TECHNOLOGY KILL
THE GOOSE THAT LAID THE
GOLDEN EGG OR SAVE IT?

New media, old media, and the #NODAPL movement

As social media caught up with the Dakota Access Pipeline resistance in the summer and autumn of 2016, the #NODAPL hashtag was birthed. For many now, #NODAPL is synonymous with the resistance by the Standing Rock Sioux Tribe to the environmental injustice of an oil pipeline being constructed upstream from their water source. Social media, along with the gathering of Indigenous "water protectors" from many Tribes around the world, was always a cornerstone in the fight against the pipeline and thus seems a fitting discussion to end this book.

Valeria Alia (2012) notes that traditional "media depictions of Indigenous peoples are steeped in the language of conquest and colonization" (p. 33). Part of the reason for this, as I have contended in this book, is due to the flawed structures of the professional and commercial media landscape, or what I have termed *deep media*. The strong gatekeeping function carried out by the mainstream media meant that the struggle by the Tribe against the Dakota Access Pipeline seemed destined for obscurity – at least where the press was concerned. As a result, perhaps the biggest surprise for those following the Standing Rock demonstrations was the impact of social media to garner worldwide attention for the Tribe's cause.

As the Tribe made clear during the interviews, they perceived clear benefits to having such a significant social media presence at Standing Rock. One of the advantages of social media – and particularly Amy Goodman's viral video – was that it spurred legacy media outlets into action, which meant that the Tribe's environmental justice concerns garnered more attention on the national and international stage – no mean feat, given that issues like these are typically ignored or minimized by the news media. In addition, some members of the Tribe noted that social media permitted their stories to be told without the institutional "filter" that they felt often distorted their perspectives, words, and actions. Finally, as Standing Rock Sioux Tribe member Phyllis Young noted, social media meant that it was no longer "your word against mine." From this perspective, social media provided a way to inform media audiences around the globe about what was actually happening at Standing Rock.

As such, the #NODAPL movement demonstrated the power of new media to transform the old media landscape. But, as with anything, nothing is black and white, and it is important to note just *how* social media shaped the demonstrations in 2016 – and particularly what it might mean for our understanding of where journalism might be headed in response. I write this final concluding chapter with some caution, for I am not a "new media" scholar who regularly engages with the numerous theories that abound within that specific area of study. Instead, I take some space here to discuss the role of new media in the #NODAPL movement specifically. While I discern some broad patterns and trends in the media landscape, I stop short of broad proclamations about the future of new media, and instead gladly let others take some of these observations further in their own work. Discussions like these, however brief, are important, for although Indigenous people across the globe are embracing social media, "Indigenous activism and social media research has yet to gain any real traction in academia" (Wilson et al., 2017, p. 1).

The shifting terrain of the media landscape at Standing Rock

Chris Hedges (2009) has written that corporations

> saturate the airwaves, the Internet, newspapers, and magazines with advertisements promoting their brands as the friendly face of the corporation.... They hold a near monopoly on all electronic and printed sources of information, A few media giants control nearly everything we read, see, and hear.
>
> *(p. 163)*

This statement still holds true in that there continues to be concentration and consolidation in the media industry, especially with growing giants like Comcast and the more established Walt Disney conglomerate. But now, the role that social media played in the struggle at Standing Rock (as well as in the MeToo, #BlackLivesMatter, #NeverAgain, and countless other movements) seems to have thrown the idea of all-encompassing dominance into question. And a large part of this confrontation of the ossified structures of old media by new media relates to decentralization of communication in the Information Age.

Ithiel de Sola Pool, discussing what he notes is related to a *soft technological determinism*, contends that "freedom is fostered when the means of communication are dispersed, decentralized, and easily available.... Central control is more likely when the means of communication are concentrated, monopolized, and

scarce" (1983, p. 5). Along these lines, Downing (2011) suggests that the proliferation of various communication channels created by "newer technologies"
might mean that concentrated ownership (of film, of the press, and of radio)
holds far less influence than in the past:

> The restrictions had been put in place to secure a diversity of content and
> opinion via a reasonable multiplicity of ownership, but that had been on
> the assumption that market entry to major media was prohibitively costly.
> Freedom of mediated communication was now in the realm of the present.
>
> *(pp. 151–152)*

This transformation, of course, seems well underway, where social media most
closely resembles a "polycentric, horizontally networked, flexible, and dynamic
structure" that has replaced the "centralized, hierarchical, strictly ruled structure
in the past" (Coban, 2016, p. xiii). An integral part of this challenge to established power structures, seen often during the Standing Rock demonstrations,
is the dissolution of what is known as *information asymmetry* – where experts and
official sources possess and wield important information as a form of power.
While many new media scholars do not believe the Internet has completely
resolved this imbalance, it is clear that at Standing Rock social media mitigated
it significantly.

One can certainly argue that the Internet, as a decentralized means of communication, served as a democratizing tool during the height of #NODAPL, for
it enabled the Standing Rock Sioux Tribe and other demonstrators to disseminate much needed information and to gain support for their cause. On myriad
social media platforms the voices of the "water protectors" provided the foundation for the growing global discussion about the pipeline through the use of
the "uniquely versatile system" of the Internet for many of the functions identified by Murdock and Golding (2004, pp. 244–245): a "distribution channel for
information…a forum for social exchange…a space of individual expression and
creativity, and an arena of social and political organization." As noted, if not for
social media and the Internet the Tribe's fight might never have been covered in
depth by traditional news media gatekeepers – although it is impossible to state
that definitively, for social media coverage was there, and it was there relatively
early in the struggle.

Old versus New Media: symbiotic, adversarial, or transformational?

Owen (2014) contends that there is a new way to think about the interconnection of what is increasingly referred to as "legacy" media and new media:

> The relationship between traditional and new media has gone *from adversar-
> ial to symbiotic*, as new media have become sources of campaign information

for professional journalists. Average citizens have become prolific providers of election-related content ranging from short reactions to campaign stories to lengthy firsthand accounts of campaigns events. Mainstream media have tacked new media features onto their digital platforms, which have become delivery systems for content that originates from websites, Twitter feeds, blogs, and citizen-produced videos. As a result, messages originating in new media increasingly set the campaign agenda.

(p. 1, emphasis added)

Although Owen writes in the context of electoral campaigns and not social movements, it is clear that new media *have* precipitated changes in the traditional media landscape across the board, with #NODAPL being a prime example. Considering the events at Standing Rock specifically, I posit that the relationship was less *symbiotic* (which suggests a mutual interdependence) and more closely represented a tense (and at times adversarial) power struggle. In this sense I agree with the idea that while conventional media produce "ideologically dominant definitions of reality, their erstwhile dominance is under severe challenge in the Internet era" (Hackett et al., 2017, p. 2).

Whether or not this new media "challenge" led to transformation of the old media system really depended on what news source one was paying attention to. For some outlets, the #NODAPL social media campaign and Amy Goodman's video of dogs biting the protesters had no effect on their coverage. No matter what fever pitch on social media was reached and no matter the stage of the protests, many newspapers in Canada refused to cover the issue in any consistent way. This overall absence of coverage was so striking due to the significant similarities between Canada and the U.S.: both countries have large oil reserves that require increased pipeline infrastructure to move them to global markets, and both countries have Indigenous populations who are seeking a seat at the nation-to-nation negotiations table. As a result of these similarities, the overall silence from the Canadian press often seemed deafening. For the more local *Bismarck Tribune* in the U.S., the coverage of the "protesters" as lawless and unjustified only seemed to harden after the video aired and when the pipeline was under construction. For the editors of that paper, it was clear that they felt they had lost the important "PR battle" (as they put it) on social media. Perhaps as a result of this perception, they appeared to take an adversarial approach to new media – one that often was manifested in turning a blind eye to the explosion of information from social media at Standing Rock that made its way around the world.

Regardless of the silence and minimization of the Tribe's concerns exhibited by many news sources, however, several mainstream outlets sat up and paid attention. In interviews, the Standing Rock Sioux Tribe and journalist Mark Trahant gave high praise to MSNBC's Lawrence O'Donnell, who reported on the need for fair treatment for the Tribe, including the need to recognize and uphold treaty rights. *The Guardian* in the UK reported on the environmental justice angle to the story consistently and tried to separate fact from fiction for its international readers. The *New York Times*, part of my framing analysis, provided perhaps the best example of a paper whose coverage of the pipeline pivoted from more economic stance (where the pipeline was viewed through the lens of how much money would be made or lost) to an environmental justice perspective that underscored the environmental and cultural needs of the Tribe. Importantly, this shift seemed to take place after Goodman's video of the dogs biting the demonstrators went viral. As such, we see (in a point made several times in this book) that social media was integral in spurring mainstream news to cover the issue more frequently and with more empathy and understanding.

Writing specifically on the role of social media in social movements, many scholars perceive an "equalizing trend" is well underway. Eren-Erdogmus and Ergun (2017) observe that "social media have proven to be strategic for initiating, organizing, and communicating social movements," in part because they can be effective catalysts for increased coverage of those movements (p. 224). Alison Anderson (2014) identifies this type of social media use as "mass self-communication," which she believes provides the opportunity for social movements to resist and challenge the status quo while also allowing significantly more voices to be heard (p. 21). In which direction the movements may go with social media is undetermined, which Anderson also sees as significant. In all, social media use at Standing Rock demonstrated that "effective use of messenger services of Facebook, Twitter, YouTube, blogs, and mobile phones became irreplaceable for revolutionary organization and mobilization in the process called 'social media revolution'" (Coban, 2016, p. xi).

In this sense, the social media presence at Standing Rock turned out to be transformational. This was not just because it precipitated a change in some of the old media institutions, although that can be seen as part of it. Instead, as many have observed, it enabled the firsthand experiences of the "water protectors" to be directly communicated to watchful audiences around the globe as part of *testimony*. Writing on the social movement in Oaxaca in 2006, Lynn Stephen (2013) notes that *testimony* is a key form of political participation in Indigenous and rural communities:

> Most basically, testimony refers to a person's account of an event or experience as delivered from the lips of a person through a speech act. It is an oral telling of a person's perception of an event. It signifies witnessing, from the Latin root *testis*, or witness.
>
> *(p. 2)*

When the different members of the Tribe talked about social media cutting through the "filter" or avoiding the "he said, she said" of the mainstream media, they seemed to refer to this important form of truth telling. For a group of people who have experienced historic injustices, as well as facing modern ones like the pipeline, telling the truth of the matter turns out to be crucial, and social media became the vehicle for this form of testimony.

Normative theories regarding the relationship between democracy and the media underscore the need for the public to have unfettered access to information and an open, accessible public space for critical perspectives. Journalism that retains its watchdog function could be complemented by – and work in tandem with – the myriad critical perspectives emanating from social media. This type of symbiosis would foster the type of civic engagement that is a hallmark of a healthy, functioning democracy. However, this research has revealed that – for Standing Rock at least – the press has not reached a fully symbiotic relationship with new media but in many ways retained its adversarial stance. Given that the phenomenon of social media shows no signs of abating (only intensifying), it is important to explore where journalism might go next in response. Will it harden its corporate stance, or will it react to the critical voices in social media by adopting more of a public service stance? These questions are outside the scope of this chapter, which remains focused on Standing Rock, but they are worth asking and exploring. The next section addresses an oft-stated idea: that new media have "killed" journalism. In so doing, it returns us to discussions earlier in this book regarding the structural weaknesses in journalism, and how it might recover in the face of the new media explosion.

Did technology kill the goose that laid the golden egg?

As noted at the beginning of this book, many political economy scholars (Couldry and Turow, 2014; McChesney, 2008; Zelizer, 2008; Chester, 2007) have written about new media's impact on journalism. McChesney (2014) believes that "the Internet has accelerated the demise of commercial journalism and made it irreversible" (p. 19). The danger, he argues, is that "now digital content can be spread instantly, at no charge, all over the world with the push of a button. The marginal cost of reproducing material is zero, nothing, nada. By free market economics, that is its legitimate price" (pp. 19–20). All of this, according to McChesney, has a deleterious impact on journalism, for if no one is willing to pay for it, a void is created that is likely to be filled by corporate interests. Zelizer (2008, p. 193) seems to agree and is specific about why she thinks this:

> In much of conventional parlance, the crisis is due primarily to the Internet providing competition to the dominant commercial news media and draining resources from the traditional journalism. This has led to an economic downturn for broadcast news and, especially, for daily newspapers, the guts of news procurement in the United States. As the Internet takes away advertisers and readers, daily newspapers lay off journalists, board up newsrooms, and prepare to join the horse-and-buggy in the annals of American history. And the marketplace has provided no economic alternative to generate the resources for journalism as we know it online so society loses. The market has spoken, for now at least. Technology killed the goose that laid the golden egg.

In this scenario, the "golden egg" is a metaphor for the fourth estate, which technology (in the form of new media) seems destined to topple. The point that

Zelizer makes is well taken, for journalism has suffered in myriad ways in the Internet era.[1] President Trump won't be around forever, and then what happens to the fortunes of the big dailies who experienced a paradoxical "bump" from news audiences captivated by the former reality TV star's antics? Put another way, when Trump fades from view, will we still pay to fervently consume professional news? No one yet knows, but political economy scholarship has documented the overall trend of declining subscriber monies for journalism well enough.

Instead of focusing on the idea of death (in that technology has "killed" traditional forms of news), perhaps we can view the media landscape through the lens of adaptation. The corporate ownership of news, as well as the commercial pressures facing newsrooms, has created a lapse in muckraking journalism, where former watchdogs are more likely to become lapdogs in relation to power structures. This void in critical perspective, it turns out, has been quickly filled by "citizen journalists"– as well as simply those who know how to wield a camera (or drone) – to bring attention to a cause or situation. In social movements the practice of first-person *testimony* through social media continues to rise, and the individual perspectives *are* critical of those in power. If journalism cannot fulfill that critical function, it will become even more sidelined from public discussions about pressing social and environmental justice issues. And when it becomes sidelined, other, more critical forms of information dissemination will fill that gap – be it the "infotainment" of John Oliver and Trevor Noah or the first-person testimony of social media platforms.

In this shifting terrain of the media landscape, there may be more questions than answers. Is it more productive to focus less on the death of journalism and more on the welcome growth of critical perspectives through social media? Should we place greater emphasis on how new media might transform traditional print journalism in ways that might benefit citizens? It depends on where you stand. When it comes to Standing Rock, however, the answer is clear: instead of seeing #NODAPL through the lens of journalism's demise, what if we saw the rebirth of participatory democracy that readily includes numerous diverse voices and viewpoints?

Sovereign nations, social media, and social movements: a "digital ecosystem"

During my interviews with them, the Standing Rock Sioux Tribe members were careful to note that the protocol for their resistance to the Dakota Access Pipeline was established through prayer and ceremony: it was to be non-violent while promoting social and environmental justice. Alia (2012), in her *New Media Nation*, notes that one way to do this is by "waging war nonviolently" through Indigenous media projects. The videos, images, and stories that came out of Standing Rock did just this: they galvanized support for the Tribe in ways that surprised just about everyone. Alia coined the phrase *new media nation* as a way to refer to a solidarity that "exists outside the control of any particular nation state, and enables its creators and users to network and engage in transcultural and transnational lobbying, and access information that might otherwise be inaccessible within state borders" (pp. 7–8). A *new media nation*, then, represents a shift in terms of who holds power, for it is marked by a significant proliferation of Indigenous news media as well as artistic and cultural productions like film and music. The "explosion" of social media coverage of the #NODAPL movement that helped to draw the attention of traditional news media thus can be seen as significantly shifting the seat of power when it came to the representation of the "water protectors."

In the broad sense of the "nationhood," the explosion of social media focused on Standing Rock seems to represent the conception of a "Fourth World." Alia (2012, p. 12[2]) explains that the phrase came about as part of a conversation between Tanzanian government official Mbuto Milando and Indigenous leader George Manuel, and was later adapted by Manuel and Posluns in *The Fourth World: An Indian Reality* in 1974. Within it is a call for Indigenous solidarity that extends beyond national borders as part of a decolonization movement (Alia, 2012). Although Manuel and Posluns were not considering the the potential of social media (as we know it today) in this vision, the #NODAPL movement has made clear that new media may be one tool in increasing a solidarity that moves more fluidly across many types of borders.[3]

Examples abound of Indigenous people using social media to aid their social justice movements. Gravante and Poma (2016) chronicle how the Spanish "Indignados" movement protesting the political and class system was undergirded/supported by new media in 2011. Speaking to the power of Indigenous-produced social media, Cherokee writer and comic artist Roy Boney, Jr. (2012) perceives an "equalizing trend" enabled by new media technology. In this trend, "technologies that were once the domain of money-laden media studios are now widely available to anyone who can afford a…computer, video camera, or mobile phone" (p. 222). This new "digital ecosystem"[4] (his words) enabled the Cherokee community to use their own language on social media platforms as well, helping to combat stereotypes and other power imbalances. What these examples reveal is how new media are changing the game when it comes to Indigenous expression.

In writing about what happened at Standing Rock in the last few months of the demonstrations, Mark Trahant (2017) observed that

> It's true that North Dakota used the power of the state this week to march into the camps along the Missouri River and round up the remaining water protectors. But think about that story: A few dozen people were arrested at camps where more than 10,000 people once said 'no' to the Dakota Access pipeline. That people power did not evaporate. Even today more people, tens of thousands, perhaps millions, are still a part of that effort via social media and through direct action.

In so stating, Trahant makes the case for the ability of social media to support a movement through gaining support ("people power") and providing a vehicle for mobilization and action. As the few previous sections of this chapter have made clear, social media was a powerful force in the demonstrations at Standing Rock. It worried corporations and governments, seeming to make it more likely that they would begin to consult with Indigenous nations regarding future developments. In Canada, it seems this already is happening, as more First Nations groups are making the news in the process of government-to-government negotiations on large infrastructure projects.

Marketing resistance: the commercial platform of social media

At this juncture, it is essential to temper what might seem like an idealistic view regarding social media's role in social movements – that is, it exists as a vehicle for testimony and self-expression in social movements that furthers democratic ends unequivocally. Mosco (2004, p. 215) makes this point well, warning us against the growing view of a "virtual utopia." The explosion of social media used in social movements, of course, can have downsides, and the Standing Rock Sioux Tribe was well aware of this. One of the disadvantages that the Tribe, as well as journalist Mark Trahant, mentioned was the rumor mill that the Internet (and social media specifically) seems particularly well suited to enable. In addition, while the Tribe received a large amount of social media attention that helped its environmental justice cause, there also were stories circulating widely on the Internet that actively hurt their reputation and their ability to combat the pipeline. In this sense, one must consider McLuhan's idea of the medium *as* the message. Speaking about the eventual reduction of social media attention paid to Standing Rock, Dreyfuss (2017) explains this well:

> the whiplash of the news cycle and the short attention spans exacerbated by the Twitterification of politics worked against those efforts. On Tuesday, the hashtag #DAPL trended nationwide for a little while, and then was eclipsed by chatter about the Academy Awards nominations. If social media and live streaming enabled the Standing Rock Sioux to amplify their protest for clean water, its speed and ceaseless flow also allowed the world to forget about them.

Although the Tribe continued to resist the pipeline and also to fight numerous court battles, the #NODAPL (and #NDAPL) social media attention quickly quieted as more issues (and hashtags) were born, and that, as Dreyfuss notes, is part of the problem.

In addition, one must also consider the *commercial* nature of social media sites and how that might impact the messages emanating from social movements. While Uldam and Vestergaard (2015) agree that social media is a potent vehicle for transformational change outside the mainstream media, they caution a utopian view of social media due to the commercialization of social media platforms. Dencik and Leistert (2015, p. 6) pinpoint the exact concern that undergirds seeing social media as a value–neutral instrument:

the continuous production of data on commercial social media platforms has both intended and unintended consequences for activists. It shapes, shifts and structures protest dynamics in contradictory ways and it simultaneously embeds protest practices within a system based on mining and monitoring activity by a number of different actors.

The idea that one must consider the role of social media as a commercial platform when employed during protests is important to consider, for of course these social media sites are provided for "free" due to the information mining that is the hallmark of Big Data. The type of structured and unstructured data generated by any online activity, including the explosion of social media seen at the height of the demonstrations at Standing Rock, is of course collected, analyzed, and sold to marketers. This recognition serves to remind us of the fact "that 'new' media technologies often present a good deal of continuity, especially in terms of corporate involvement, commercialization and commodification" (Wasko, 2014, p. 267).

If there is a question regarding just how critical perspectives become part of the marketing machine one need simply look to the advertising industry – the same ones that created PR disasters like the 2017 Pepsi/Kendall Jenner video (which used the unrest associated with various social justice protests to market the soda) or Diesel Apparel's "Global Warming Ready" campaign in 2007 (which used images of wealthy, thin, young white people "surviving" catastrophic climate change in style around the world). As such, the digital world is not much different: as Mosco has it, the persistent myths of a utopian cyberspace seem to ignore the crucial fact that "digitization takes place along with the process of commodification" (p. 215). As indicated by these and myriad other examples, capitalism (and its attendant consumer culture) always find ways to commodify resistance, and thus is worth exploring in the context of social movements.

Regardless of the pure commercialism of social media platforms, those used most often in social protest (such as Twitter and the currently maligned Facebook) make clear that social media still fills an important gap left by journalism. Leistert (2015, p. 35) writes that

> For protests against established power structures, against crude social injustice in the twenty-first century and, in very general terms, for a life in dignity, the media technologies of choice are corporate social media platforms precisely because they have become what traditional mass media has ceased to be: a way to inform the general population about relevant societal issues.

As such, while one needs to consider the commercial motivations of social media sites, it also is the case that for those who gathered at Standing Rock, these vehicles for self-expression were invaluable.

Conclusion: a new direction for coverage of social justice issues

Normative theories regarding the relationship between democracy and the media underscore the need for the public to have unfettered access to information and a healthy, open space for critical perspectives. A strong news media landscape that retains its watchdog function could complement the myriad critical perspectives emanating from social media – in the process fostering the civic engagement that is a hallmark of a healthy, functioning democracy. This book has revealed, among other things, that the press has not reached a fully symbiotic relationship with new media but in many ways retains its adversarial stance. In her provocatively titled essay "Public Journalism Is a Joke," Faina (2013) provides suggestions as to how to rehabilitate journalism, including that journalists should recognize the different values at play in a story, including their own personal ones as well as those held by their parent institution. He writes that

> incorporating competing values into news coverage recognizes that any action taken, whether by voting for a specific politician, supporting a course of action, or introducing a certain policy, is going to have a range of consequences. Journalists would be wise to spend time focusing on what these consequences are so that individuals can better gauge their decisions.[5]
>
> *(p. 544)*

In addition, journalists should focus on the important issues undergirding any story: "journalists should work to uncover overarching themes and problems.... This will better enable the public to use the news to enact a more robust and effective public life" (p. 554). Finally, and perhaps most germane to this research, Faina argues that journalists should always be cognizant of how they *frame* their stories. He particularly calls out the Conflict frame as problematic because a focus on acrimonious public fights tends to inhibit the public's desire to engage with an issue. Instead, stories should be framed "in terms of [their] implications for a community rather than focus[ing] on who will benefit politically from a given situation" (p. 545). Seen from this perspective, stories should lean more towards community concern, which more easily aligns with addressing social justice issues like those at Standing Rock.

But there is more to consider than just this. There are very few Indigenous journalists hired at the larger national dailies, and so the Indigenous perspective often is omitted from stories involving Native Americans (Hernandez, 2012). With the lack of perspective often comes lack of context, which includes the important history and background needed for media audiences to understand and take action on an issue. Kemper (2010) broadly identifies just why having Indigenous journalists is so important: "Tribal journalists have used the printing press and subsequent technologies since the early nineteenth century to tell... stories of survival, persistence, resilience, and successes.... They also have used these technologies to correct misrepresentations about indigenous peoples" (p. 5). The inclusion of more diverse and experienced voices is thus needed in the media landscape: if mainstream media has not included all voices, then "mass self-communication" using digital media platforms becomes even more important in social justice movements involving Indigenous people.

In all, it became impossible to ignore the power of first-person "testimony" through social media at Standing Rock in the autumn of 2016. An environmental justice issue that seemed destined to receive the same treatment – silence – from the mainstream press was suddenly, at least for some newspapers, front-page news. Part of the reason for this was, of course, the viral videos spreading like wildfire that the press ignored only at their peril. In the end, the pipeline did get built and the oil currently flows through the pipes in a turn of events that ultimately feels familiar. The oil companies and the state wanted the oil, and in the end, that is what they got. But at what cost? As Duarte (2017, p. 10) observes:

> The expression of strong mediatised Indigenous voices does not ensure that dominant government authorities will accordingly listen or act in a just manner. Effective political participation for Indigenous peoples vis-à-vis nation-states still requires significant structural changes. Meanwhile, Indigenous activists, entrepreneurs, educators, and many other leaders must effectively and strategically push however they can from whatever digital/social/political position they hold, the embodiments of decolonisation and perpetual performers of radical change.

To say that the efforts of the Standing Rock Sioux Tribe were not successful because their ultimate goal was not met is a conclusion I have heard often while writing this book. And while the Tribe noted that they had lost much-needed revenue during the demonstrations as well as the all-important access to clean water (if the pipeline catastrophically breaks above their water source), to categorize the movement as a failure would be a misrepresentation of what the Tribe accomplished. First, the world listened intently to the voices of the "water protectors" and provided support for the movement. In addition, the Tribe also effectively changed the contours of the issue by promoting the issue through the lens of environmental justice and sparking enduring conversations about climate change, pollution, and the need to focus more on renewable energy. The Tribe also gained the support of numerous environmental and political groups as well as other social movements and veterans of the U.S. military. Finally, the Tribe noted to me that they deeply valued the historic level of solidarity and communication with other Indigenous groups around the globe. They saw those ties as being important connections to have going into the future.

In sum, what Wilson et al. (2017, p. 3) write about social media in general also seems true for the movement at Standing Rock:

Social media is now undoubtedly a great force for political organisation and transformation allowing for the amplification of Indigenous voices. These voices...reaffirm Indigenous sovereignty and connections to ancestral homelands. Indigenous people indeed are reterritorialising social media and as always will continue to Rise Up.

The resistance to the Dakota Access Pipeline taught the world to pay attention: to environmental injustice, to Indigenous voices, and to the environmental degradation we currently are witnessing. Thus, what Alia (2012) has observed seems to hold true: "Indigenous people are changing the media, and with it, the world" (p. 172).

Notes

1 Intriguingly, Hernandez (2012) contends that the "economic tsunami" that rocked U. S. newspapers in 2008 and 2009 did not impact Indigenous journalists very much. The reason? "There are simply not enough American Indian journalists working at mainstream dailies to feel the impact" (p. 227).
2 Alia, pulling from McFarlane (1993, pp. 160–161).
3 Alia recognizes the difficulty inherent in the category of "Indigenous peoples" in that such a reference does not recognize individual sovereignty or nationhood status.
4 Boney is specifically referring to the ability of people in the Cherokee community to use their own language when communicating on new media technology.
5 For most of these principles, Faina cites Merritt, D. (1998). *Public journalism and public life: Why telling the news is not enough*. Mahwah, NJ: Lawrence Erlbaum.

References

Alia, V. (2012). *The new media nation: Indigenous peoples and global communication*. New York: Berghahn Books.

Anderson, A. (2014). *Media, environment and the network society*. New York: Palgrave Macmillan.

Boney, R., Jr. (2012). Cherokeespace.com: Native social networking. In M. Carstarphen & J. Sanchez (Eds.), *American Indians and the mass media*. Norman, OK: University of Oklahoma Press.

Chester, J. (2007). *Digital destiny: New media and the future of democracy*. New York: New Press. Distributed by W.W. Norton.

Çoban, B. (2016). *Social media and social movements: The transformation of communication patterns*. Lanham, MD: Lexington Books.

Couldry, N., & Turow, J. (2014). Advertising, big data, and the clearance of the public realm: Marketers' new approaches to the content subsidy. *International Journal of Communication, 8*, 1710–1726.

de Sola Pool, I. (1983). *Technologies of freedom*. Cambridge, MA: Belknap Press.

Dencik, L., & Leistert, O. (2015). *Critical perspectives on social media and protest: Between control and emancipation*. New York: Rowman & Littlefield International.

Downing, J. (2011). Media ownership, concentration, and control: The evolution of debate. In J. Wasko, G. Murdock, & H. Sousa (Eds.), *The handbook of political economy of communications* (pp. 140–168). Malden, MA: Wiley-Blackwell.

Dreyfuss, E. (2017, January 24). Social media made the world care about Standing Rock – and it helped it forget. *Wired.* Retrieved from https://www.wired.com/2017/01/social-media-made-world-care-standing-rock-helped-forget/

Duarte, M. (2017). Connected activism: Indigenous uses of social media for shaping political change. *Australasian Journal of Information Systems, 21.* 10.3127/ajis.v21i0.1525

Eren-Erdogmux, I., & Ergun, S. (2017). The impact of social media on social movements. In S. Gordon (Ed.), *Online Communities as Agents of Change and Social Movements* (pp. 224–252). Hershey, PA: IGI Global, Information Science Reference.

Faina, J. (2013). Public journalism is a joke: The case for Jon Stewart and Stephen Colbert. *Journalism, 14*(4), 541–555. 10.1177/1464884912448899

Gravante, T., & Poma, A. (2016). New media and the empowerment in the Indignados. *Social media and social movements* (pp. 19–36). New York: Lexington Books.

Hackett, R. A., Forde, S., Gunster, S., Foxwell-Norton, K., Hackett, R. A., Forde, S., & Foxwell-Norton, K. (2017). *Journalism and climate crisis: Public engagement, media alternatives.* New York: Routledge.

Hedges, C. (2009). *Empire of illusion: The end of literacy and the triumph of spectacle.* New York: Nation Books.

Hernandez, J. A. (2012). Native Americans in the twenty-first century newsroom. In M. Carstarphen & J. Sanchez (Eds.), *American Indians and the Mass Media* (pp. 227–231). Norman, OK: University of Oklahoma Press.

Kemper, K. R. (2010). Who speaks for indigenous peoples? tribal journalists, rhetorical sovereignty, and freedom of expression. *Journalism & Mass Communication Monographs, 12*(1), 3–58. 10.1177/152263791001200101.

Leistert, O. (2015). The revolution will not be liked. In L. Dencik & O. Leistert (Eds.), *Critical perspectives on social media and protest* (pp. 35–52). New York: Rowman & Littlefield.

McChesney, R. W. (2008). *The political economy of media: Enduring issues, emerging dilemmas.* New York: Monthly Review Press.

McChesney, R. (2014). The struggle for democratic media: Lessons from the north and from the left; In C. Martens, E. Vivares, & R. McChesney (Eds.), *The international political economy of communication: Media and power in South America* (pp. 11–30). London: Palgrave Macmillan.

Mosco, V. (2004). Capitalism's Chernobyl? From ground zero to cyberspace and back again. In A. Calabrese & C. Sparks (Eds.), *Toward a political economy of culture* (pp. 211–243). New York: Rowman & Littlefield.

Murdock, G. & Golding, P. (2004). Dismantling the digital divide. In A. Calabrese and C. Sparks (Eds.), *Toward a political economy of culture* (pp. 244–260). New York: Rowman & Littlefield .

Owen, D. (2014). New media and political campaigns. In K. Kenski & K. Jamieson (Eds.), *The Oxford handbook of political communication* (pp. 1–21). New York: Oxford University Press.

Stephen, L. (2013). *We are the face of Oaxaca: Testimony and social movements.* Durham, NC: Duke University Press.

Trahant, M. (2017, February 24). Water is life: The story of Standing Rock won't go away. *Yes! Magazine.* Retrieved from http://www.yesmagazine.org/people-power/water-is-life-the-story-of-standing-rock-wont-go-away-20170224

Uldam, J., & Vestergaard, A. (2015). *Civic engagement and social media: Political participation beyond protest.* New York : Palgrave Macmillan.

Wasko, J. (2014). The study of the political economy of the media in the twenty-first century. *International Journal of Media & Cultural Politics*, *10*(3), 251–279.

Wilson, A., Carlson, B., & Sciascia, A. (2017). Reterritorialising social media: Indigenous people rise up. *Australasian Journal of Information Systems*, *21*. 10.3127/ajis.v21i0.1591.

Zelizer, B. (2008). *Explorations in communication and history*. New York: Routledge.

APPENDIX A

Standing Rock interview questions: "Journalism and the Dakota Access Pipeline" project

Questions during each semi-structured interview include:

1. What is your perspective on local and national media coverage in terms of <u>quantity</u> (that is, was it covered often enough, and did the issue receive enough attention? Do you think the media are still paying attention now?)

2. What are your thoughts on the <u>quality</u> of the coverage you received, especially in terms of how your protest activities were represented?

3. Do you think Amy Goodman's video on Labor Day that documented the pipeline's use of force, including dogs and pepper spray, had an impact on how your struggle was represented to people in the U.S. and around the world? How and why?

4. [*Related to Question 3*]: How do you perceive of the role of social media in your struggle: as an effective tool for you, as a hindrance to your cause? And, do you think that it shaped the local, national, or international coverage you received?

5. Can you share your thoughts and perspectives on the relationship between oil, water, and our environment?

6. Please provide your perspective on how you think local or national politics shaped the NDAPL movement: you are welcome to take this in any direction you like, including discussions of local/state politics, Presidents Obama and Trump, or anything else you would like to share.

7. Finally, are there any other perspectives or thoughts you would like to share, anything I have not asked about here?

Please note that because this is a semi-structured interview, other questions may arise as a way to explore additional information related to participants' unique responses.

APPENDIX B

Interview questions for Mark Trahant: "Journalism and the Dakota Access Pipeline" project

Questions during each semi-structured interview include:

1. What is your perspective on local and national media coverage of the #NODAPL movement in terms of <u>quantity</u>: that is, was it covered often enough, and did the issue receive enough attention? Do you think the media are still paying attention now?
2. What are your thoughts on the <u>quality</u> of the coverage the movement received, especially in terms of how the demonstrations were represented?
3. Do you think Amy Goodman's video on Labor Day that documented the pipeline's use of force, including dogs and pepper spray, had an impact on how the #NODAPL struggle was represented to people in the U.S. and around the world? How and why?
4. [*Related to Question 3*]: How do you perceive of the role of social media in the movement: as an effective tool, as a hindrance? And, do you think that it shaped the local, national, or international coverage?
5. Can you share your thoughts and perspectives on the relationship between oil, water, and our environment?
6. Please provide your perspective on how you think local or national politics shaped the #NODAPL movement: you are welcome to take this in any direction you like, including discussions of local/state politics, Presidents Obama and Trump, or anything else you would like to share.
7. Finally, are there any other perspectives or thoughts you would like to share, anything I have not asked about here?

Please note that because this is a semi-structured interview, other questions may arise as a way to explore additional information related to participants' unique responses.

INDEX